Buckle Down™

to the
Common Core
State Standards

Mathematics
Grade 3

This book belongs to: _____

Buckle Down

Helping your schoolhouse meet the standards of the statehouse™

ISBN 978-0-7836-7985-3

1CCUS03MM01

8 9 10

Cover Image: Colorful marbles. © Corbis/Photolibrary

Triumph Learning® 136 Madison Avenue, 7th Floor, New York, NY 10016

© 2011 Triumph Learning, LLC
Buckle Down is an imprint of Triumph Learning®

Frequently Asked Questions about the Common Core State Standards

What are the Common Core State Standards?

The Common Core State Standards for mathematics and English language arts, grades K–12, are a set of shared goals and expectations for the knowledge and skills that will help students succeed. They allow students to understand what is expected of them and to become progressively more proficient in understanding and using mathematics and English language arts. Teachers will be better equipped to know exactly what they must do to help students learn and to establish individualized benchmarks for them.

Will the Common Core State Standards tell teachers how and what to teach?

No. Because the best understanding of what works in the classroom comes from teachers, these standards will establish *what* students need to learn, but they will not dictate *how* teachers should teach. Instead, schools and teachers will decide how best to help students reach the standards.

What will the Common Core State Standards mean for students?

The standards will provide a clear, consistent understanding of what is expected of student learning across the country. Common standards will not prevent different levels of achievement among students, but they will ensure more consistent exposure to materials and learning experiences through curriculum, instruction, teacher preparation, and other supports for student learning. These standards will help give students the knowledge and skills they need to succeed in college and careers.

Do the Common Core State Standards focus on skills and content knowledge?

Yes. The Common Core State Standards recognize that both content and skills are important. They require rigorous content and application of knowledge through higher-order thinking skills. The English language arts standards require certain critical content for all students, including classic myths and stories from around the world, America's founding documents, foundational American literature, and Shakespeare. The remaining crucial decisions about content are left to state and local determination. In addition to content coverage, the Common Core State Standards require that students systematically acquire knowledge of literature and other disciplines through reading, writing, speaking, and listening.

In mathematics, the Common Core State Standards lay a solid foundation in whole numbers, addition, subtraction, multiplication, division, fractions, and decimals. Together, these elements support a student's ability to learn and apply more demanding math concepts and procedures.

The Common Core State Standards require that students develop a depth of understanding and ability to apply English language arts and mathematics to novel situations, as college students and employees regularly do.

Will common assessments be developed?

It will be up to the states: some states plan to come together voluntarily to develop a common assessment system. A state-led consortium on assessment would be grounded in the following principles: allowing for comparison across students, schools, districts, states and nations; creating economies of scale; providing information and supporting more effective teaching and learning; and preparing students for college and careers.

TABLE OF CONTENTS

			Common Core State Standards

			Common Core State Standards

			Common Core State Standards

Common Core State Standards listed:
- Lesson 28 — 3.MD.7.d
- Lesson 29 — 3.MD.8
- Lesson 30 — 3.MD.3
- Lesson 31 — 3.MD.3
- Lesson 32 — 3.MD.4
- Lesson 33 — 3.G.1
- Lesson 34 — 3.G.1
- Lesson 35 — 3.G.2

To the Teacher:

Standard numbers are listed for each lesson in the table of contents. The numbers in the shaded gray bar that runs across the tops of the pages in the workbook indicate the standards for a given page (see example to the left).

Introduction

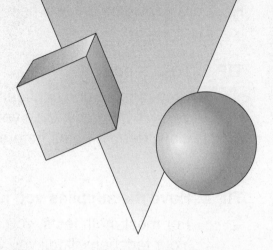

How much math do you know? You probably know a lot—even if it doesn't always seem that way. You have been using math for years. Every year, you add more to what you already know.

Learning to solve math problems is a lot like learning to play a new video game. You need to know what steps to take to stay in the game and come up with the right answers. Math can be a lot of fun if you know what you are doing. How do you get better at math? You practice. The more you practice, the better you will get.

Test-Taking Tips

Here are a few tips that will help you on test day.

TIP 1: Take it easy.

Stay relaxed and confident. Because you've practiced these problems, you will be ready to do your best on almost any math test. Take a few slow, deep breaths before you begin the test.

TIP 2: Have the supplies you need.

For most math tests, you will need two sharp pencils and an eraser. Your teacher will tell you whether you need anything else.

TIP 3: Read the questions more than once.

Every question is different. Some questions are more difficult than others. If you need to, read a question more than once. This will help you make a plan for solving the question.

TIP 4: Learn to "plug in" answers to multiple-choice items.

When do you "plug in"? You should "plug in" whenever your answer is different from all of the answer choices or you can't come up with an answer. Plug each answer choice into the problem and find the one that makes sense. (You can also think of this as "working backward.")

TIP 5: Answer open-ended items completely.

When answering short-response and extended-response items, show all your work to receive as many points as possible. Write neatly enough so that your calculations will be easy to follow. Make sure your answer is clearly marked.

TIP 6: Use all the test time.

Work on the test until you are told to stop. If you finish early, go back through the test and double-check your answers. You just might increase your score on the test by finding and fixing any errors you might have made.

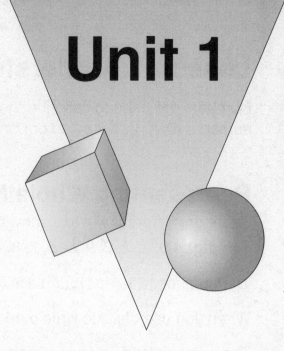

Unit 1

Number and Operations in Base Ten

Numbers are used for just about everything. You might use numbers when you put candles on a birthday cake or keep score at a baseball game or make a pitcher of lemonade. You might use large numbers to tell how far it is from your house to New York City. You might use small numbers to tell the length of your little finger.

In this unit, you will do all kinds of things with numbers—from large numbers such as 9999 to small numbers such as 1. You will round, estimate, add, and subtract numbers. You will work with patterns and properties of addition. You will also learn how to solve story problems.

Lesson 1: Understanding Place Value

Numbers are used to describe "how many" of something there are. Some numbers (0, 1, 2, 3, and so on) are known as **whole numbers**.

Representing Whole Numbers

There are different ways of showing whole numbers.

The **digits** 0, 1, 2, 3, 4, 5, 6, 7, 8, and 9 are used to write whole numbers.

The number 1236 has four digits (1, 2, 3, and 6).

When you use digits to write a number, you write the number in **standard form**.

The number 1236 is written in standard form.

When you use words to write numbers, you write the numbers as you would say them out loud.

The **number name** of 1236 is: one thousand two hundred thirty-six

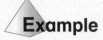

Example

How many digits does the number 49 have?

4 and 9 are digits.

The number 49 has two digits.

Example

What is the number name for 758?

Use words. Write 758 the way it sounds when you read it.

The number name for 758 is seven hundred fifty-eight.

CCSS: 3.NBT.1

Place Value

Place value helps you understand the value of each digit in a number. Each digit has a value based on its position in the number.

This place-value table shows the value of each digit in the number 5241.

Thousands	Hundreds	Tens	Ones
5	2	4	1

5000   200   40   1

Numbers written in **expanded form** show the value of each digit.
5241 = 5000 + 200 + 40 + 1

 **Example**

How is 3608 written in expanded form?

Write the number in a place-value table.

Thousands	Hundreds	Tens	Ones
3	6	0	8

Notice that there are no tens. So, a 0 (zero) is written in that place. Zeros are not shown in expanded form.

In expanded form, 3608 = 3000 + 600 + 8.

Practice

1. How is seven thousand, one hundred sixty written in standard form?

   _____

2. How is four thousand, eighty-four written in standard form? _____

3. How is eight hundred five written in expanded form? _____

4. How is 9682 written in expanded form?

_____

5. Write the number two thousand three hundred nine in the following place-value table.

Thousands	Hundreds	Tens	Ones

How is the number from the place-value table written in standard form?

_____

6. In the following place-value table, write the number that has a 4 in the thousands place, a 2 in the hundreds place, a 7 in the tens place, and a 3 in the ones place.

Thousands	Hundreds	Tens	Ones

What is the number name for the number in the place-value table?

_____

7. Circle the number that has a 7 in the hundreds place.

　　7349　　　　　8720　　　　　9673

8. Circle the number that has an 8 in the ones place.

　　383　　　　　148　　　　　844

9. Circle the number that has a 0 in the tens place.

　　2092　　　　　5301　　　　　4610

CCSS: 3.NBT.1

10. How is 700 + 80 + 4 written in standard form?

    A. 7840

    B. 7480

    C. 784

    D. 748

11. How is 4607 written in expanded form?

    A. 400 + 60 + 7

    B. 4000 + 600 + 7

    C. 4000 + 600 + 70

    D. 4000 + 600 + 60 + 7

Directions: Use the following place-value table to answer questions 12 and 13.

Thousands	Hundreds	Tens	Ones
7	5	2	8

12. What is the value of the 2?

    A. 2

    B. 20

    C. 200

    D. 2000

13. What is the value of the 7?

    A. 7

    B. 70

    C. 700

    D. 7000

14. Jack wrote a number using the clues below.

- The number has three digits.
- There is a 5 in the hundreds place.
- There is a 7 in the tens place.
- The digit in the ones place is 2 less than the digit in the hundreds place.

What is Jack's number? _____

Explain how you found the digit in the ones place.

_____

_____

Lesson 2: Properties of Addition

Properties are rules that help you work with numbers. Understanding the following properties of addition will help you with adding numbers.

Commutative Property of Addition

This rule says that the **order** in which you add two numbers **does not change the sum**.

$$4 + 2 = 2 + 4$$
$$6 = 6$$

Example

What number can replace the ☐ to make the number sentence true?

$$43 + 20 = 20 + ☐$$

According to the commutative property, the order of the numbers you are adding does not change the sum. So, 43 + 20 is equal to 20 + 43.

The number 43 makes the number sentence true. 43 + 20 = 20 + 43

Associative Property of Addition

This rule says that when you add three numbers, it doesn't matter which two numbers you add first. The **grouping** of the numbers **does not change the sum**.

$$(3 + 6) + 7 = 3 + (6 + 7)$$

Parentheses ( ) can be used to show which two numbers you add first.

$$9 + 7 = 3 + 13$$
$$16 = 16$$

CCSS: 3.NBT.2

 Example

Find the missing number in the number sentence below.

$$(12 + 5) + 15 = 12 + (\boxed{} + 15)$$

According to the associative property, the grouping of the numbers does not change the sum. The number sentence is $12 + 5 + 15$ no matter which way the numbers are grouped.

The missing number is 5: $(12 + 5) + 15 = 12 + (5 + 15)$

Identity Property of Addition

This rule says that when you **add 0 and any number**, the sum is **that same number**.

$$8 + 0 = 8 \qquad\qquad 0 + 58 = 58$$

Example

What number can replace the $\boxed{}$ to make the number sentence true?

$$47 + \boxed{} = 47$$

According to the identity property, the sum of any number and zero is that number. The only number you can add to 47 to get a sum of 47 is 0.

The number 0 makes the number sentence true. $47 + 0 = 47$

Example

What property does the number sentence show?

$$0 + 99 = 99 + 0$$

The number sentence on either side of the equal sign has the same numbers, but they are in a different order. The commutative property states that you can add numbers in any order and get the same sum.

The number sentence $0 + 99 = 99 + 0$ shows the commutative property.

Practice

Directions: For questions 1 through 6, use the properties of addition to fill in the missing numbers.

1. 6 + _____ = 6

2. (8 + _____) + 12 = _____ + (9 + _____)

3. 12 + _____ = 3 + 12

4. 5 + (11 + _____) = (_____ + _____) + 2

5. 7 + _____ = 9 + 7

6. _____ + 0 = 22

Directions: For questions 7 through 12, use the properties of addition to fill in the missing numbers. Then write the name of the property you used.

7. _____ + 6 = 6 + 8 _____

8. (_____ + 9) + 2 = 7 + (_____ + _____) _____

9. _____ + (3 + _____) = (4 + _____) + 2 _____

10. 8 + _____ = 8 _____

11. 6 + 4 = 4 + _____ _____

12. _____ + 0 = 50 _____

CCSS: 3.NBT.2

13. What is the missing number?

$\boxed{} + 5 = 5 + 7$

A. 2

B. 7

C. 8

D. 17

15. What is the missing number?

$75 + \boxed{} = 75$

A.   0

B.   1

C. 75

D. 150

14. Which shows the associative property?

A. $7 + 0 = 7$

B. $3 + 4 = 7$

C. $3 + 4 = 4 + 3$

D. $(2 + 1) + 4 = 2 + (1 + 4)$

16. Which shows the commutative property?

A. $2 + 3 = 5$

B. $2 + 0 = 2$

C. $2 + 3 = 3 + 2$

D. $(2 + 1) + 3 = 2 + (1 + 3)$

17. Add the following three numbers in the two different ways given. Be sure to show your work.

$(8 + 14) + 6$ $\qquad\qquad$ $8 + (14 + 6)$

Was the sum the same both ways? _____

Which way of adding did you find easier? Explain why.

_____

_____

_____

Lesson 3: Addition Patterns

When you look at an addition table, you can see patterns in the rows and columns. Some of these patterns can be explained by the properties of addition.

+	0	1	2	3	4	5	6	7	8	9
0	0	1	2	3	4	5	6	7	8	9
1	1	2	3	4	5	6	7	8	9	10
2	2	3	4	5	6	7	8	9	10	11
3	3	4	5	6	7	8	9	10	11	12
4	4	5	6	7	8	9	10	11	12	13
5	5	6	7	8	9	10	11	12	13	14
6	6	7	8	9	10	11	12	13	14	15
7	7	8	9	10	11	12	13	14	15	16
8	8	9	10	11	12	13	14	15	16	17
9	9	10	11	12	13	14	15	16	17	18

Example

The first unshaded column shows the sum of 0 and each number in the shaded column. The numbers in the first unshaded column are the same as the numbers in the shaded column. Which property of addition explains this pattern?

Pick two numbers: for example, 0 and 8. The table shows that $0 + 8 = 8$.

Think of the properties of addition. The identity property says that when you add 0 and any number, the answer will be equal to that number.

The identity property explains why the numbers in the first unshaded column are the same as the numbers in the shaded column.

CCSS: 3.OA.9

Example

If you draw a diagonal from the upper left to the lower right of an addition table, the numbers on either side of the diagonal match. Which property of addition explains this pattern?

Pick two numbers: for example, 2 and 3. The table shows that there is a matching sum of 5 on either side of the diagonal whether you add $2 + 3$ or $3 + 2$.

+	0	1	2	3	4
0	0	1	2	3	4
1	1	2	3	4	5
2	2	3	4	5	6
3	3	4	⑤	6	7
4	4	5	6	7	8

+	0	1	2	3	4
0	0	1	2	3	4
1	1	2	3	4	5
2	2	3	4	⑤	6
3	3	4	5	6	7
4	4	5	6	7	8

Think of the properties of addition. The commutative property says that when you change the order of the numbers to be added, the sum stays the same.

The commutative property explains why the numbers on either side of the diagonal are the same.

Even and Odd Sums

When you add numbers, you can know if the answer will be even or odd.

If you add two even numbers, the sum will be even.　　$2 + 4 = 6$　$8 + 6 = 14$

If you add two odd numbers, the sum will be even.　　$3 + 5 = 8$　$9 + 1 = 10$

If you add an even number and an odd number, the sum will be odd.　　$2 + 5 = 7$　$8 + 5 = 13$

Practice

Directions: For questions 1 through 3, use the part of the addition table shown below.

+	0	1	2	3	4	5	6	7	8	9
0	0	1	2	3	4	5	6	7	8	9
1	1	2	3	4	5	6	7	8	9	10
2	2	3	4	5	6	7	8	9	10	11
3	3	4	5	6	7	8	9	10	11	12
4	4	5	6	7	8	9	10	11	12	13

1. Look at the numbers ringed in the addition table.

   What do you notice about the ringed numbers?

   _____

2. Complete the number sentences.

   6 + _____ = 6

   5 + _____ = 6

   4 + _____ = 6

   3 + _____ = 6

   2 + _____ = 6

   What pattern do you notice in the number sentences?

   _____

3. How do your answers to question 2 relate to the addition table?

   _____

CCSS: 3.OA.9

4. Look at the numbers along the diagonals.

+	0	1	2	3	4	5	6	7	8	9
0	0	1	2	3	4	5	6	7	8	9
1	1	2	3	4	5	6	7	8	9	10
2	2	3	4	5	6	7	8	9	10	11
3	3	4	5	6	7	8	9	10	11	12
4	4	5	6	7	8	9	10	11	12	13

Write the numbers along the first diagonal. _____

Write the numbers along the second diagonal. _____

What pattern do you see in the numbers along the diagonals? _____

5. Sara added a number to 53. The sum was even. Which could be the number that Sara added?

   A. 16

   B. 30

   C. 46

   D. 49

6. Which property best explains why the row for 2 and the column for 2 in the addition table have the same numbers?

   A. associative property of addition

   B. commutative property of addition

   C. distributive property

   D. identity property

7. If you add one even number and one odd number, will the answer be even or odd?

   _____

   Give 3 examples to support your answer.

Lesson 4: Rounding 10s and 100s

Rounding a number to the nearest ten (10) means finding the ten (10, 20, 30, 40, and so on) that the number is closest to. Rounding a number to the nearest hundred (100, 200, 300, 400, and so on) means finding the hundred that the number is closest to.

You can use a number line to help you round numbers.

Example

Round 43 to the nearest ten.

Find 43 on the number line.

43 is between 40 and 50.

43 is 3 units from 40. It is 7 units from 50.

43 is closer to 40 than to 50.

43 rounded to the nearest ten is 40.

Another way to round numbers is to use place value. To round any whole number:

- **Circle** the digit in the place that you are **rounding to**.

- **Underline** the digit in the place **to the right** of the circled digit.

- The circled digit will either stay the same or increase by 1.
  If the underlined digit is **less than 5**, the circled digit **stays the same**.
  If the underlined digit is **greater than or equal to 5**, the circled digit **increases by 1**.

- Write **a zero** as a place holder for **each digit to the right** of the circled digit.

Example

Round 86 to the nearest ten.

You are rounding to the tens place, so circle the digit in the tens place, 8. The 6 is to the right of the 8, so underline it.

⑧ 6̲

Because 6 is greater than 5, the 8 rounds to 9. Write a zero as a placeholder in the ones place.

86 rounded to the nearest ten is 90.

 Example

Round 629 to the nearest hundred.

You are rounding to the hundreds place, so circle the digit in the hundreds place, 6. The 2 is to the right of the 6, so underline it.

⑥ 2̲ 9

Because 2 is less than 5, the circled digit, 6, stays the same. Write zeros as placeholders in the tens and ones places.

629 rounded to the nearest hundred is 600.

 Example

Round 317 to the nearest ten.

You are rounding to the tens place, so circle the digit in the tens place, 1. The 7 is to the right of the 1, so underline it.

3 ① 7̲

Because 7 is greater than 5, the 1 rounds to 2. Write a zero as a placeholder in the ones place. The digit before the rounded place stays the same.

317 rounded to the nearest ten is 320.

You can use rounding as a way to estimate the answer to a math problem. Estimation is a way of finding a number that is close to the exact answer. You can use estimation to check if your exact answer is reasonable (makes sense). You can also use estimation when an exact answer is not needed.

To estimate the answer to a problem, first round each of its numbers to the nearest 10 or 100. Then find the sum or difference of the rounded numbers.

 Example

Use rounding to estimate the following sum.

$$\begin{array}{r} 76 \\ + 39 \\ \hline \end{array}$$

In this example, round the numbers to the nearest ten.
76 rounds to 80. 39 rounds to 40.

Add the rounded numbers.

$$\begin{array}{r} 80 \\ + 40 \\ \hline 120 \end{array}$$

The estimated sum is 120.

Example

Use rounding to estimate the following difference.

$$\begin{array}{r} 538 \\ - 162 \\ \hline \end{array}$$

Round each number to the greatest place.
The greatest place is the hundreds place.
538 rounds to 500. 162 rounds to 200.

Subtract the rounded numbers.

$$\begin{array}{r} 500 \\ - 200 \\ \hline 300 \end{array}$$

The estimated difference is 300.

Practice

Directions: For questions 1 through 7, round each number to the given place.

1. Round 27 to the nearest ten. _____

2. Round 114 to the nearest hundred. _____

3. Round 151 to the nearest hundred. _____

4. Round 85 to the nearest ten. _____

5. Round 33 to the nearest ten. _____

6. Round 489 to the nearest hundred. _____

7. Round 826 to the nearest ten. _____

Directions: For questions 8 through 14, round each number to the nearest ten. Then round each number to the nearest hundred.

8. 138 _____ _____

9. 256 _____ _____

10. 574 _____ _____

11. 384 _____ _____

12. 419 _____ _____

13. 321 _____ _____

14. 109 _____ _____

Directions: For questions 15 through 26, use rounding to estimate the sum or difference.

15. $\begin{array}{r} 23 \\ + 56 \\ \hline \end{array}$

16. $\begin{array}{r} 81 \\ - 47 \\ \hline \end{array}$

17. $\begin{array}{r} 172 \\ + 283 \\ \hline \end{array}$

18. $\begin{array}{r} 94 \\ + 47 \\ \hline \end{array}$

19. $\begin{array}{r} 583 \\ - 516 \\ \hline \end{array}$

20. $\begin{array}{r} 61 \\ - 49 \\ \hline \end{array}$

21. $\begin{array}{r} 318 \\ - 193 \\ \hline \end{array}$

22. $\begin{array}{r} 226 \\ + 215 \\ \hline \end{array}$

23. $\begin{array}{r} 432 \\ - 197 \\ \hline \end{array}$

24. $\begin{array}{r} 127 \\ + 184 \\ \hline \end{array}$

25. $\begin{array}{r} 82 \\ - 15 \\ \hline \end{array}$

26. $\begin{array}{r} 165 \\ - 132 \\ \hline \end{array}$

27. Which is the best estimate for the difference of 79 and 23?

    A. 50

    B. 60

    C. 90

    D. 100

28. A chef ordered 175 eggs. To the nearest ten, how many eggs did the chef order?

    A. 100

    B. 170

    C. 180

    D. 200

29. Which is the best estimate for the sum of 376 and 209?

    A. 100

    B. 200

    C. 500

    D. 600

30. A grey whale is about 46 feet long. To the nearest ten feet, about how long is the gray whale?

    A. 30

    B. 40

    C. 50

    D. 60

31. Simon is given the following problem to solve.

    $$
    \begin{array}{r}
    324 \\
    + 619 \\
    \hline
    \end{array}
    $$

    Explain how Simon could use rounding to estimate the sum.

    _____

    _____

    Estimate the sum.

    What is the estimated sum? _____

CCSS: 3.OA.8, 3.NBT.2

Lesson 5: Adding Whole Numbers

When you want to find how many of something there are in all, you **add (+)**. The numbers that you add are **addends**. The answer when you add is the **sum**.

$$7 + 8 = 15$$

addends   sum

$7 + 8 = 15$ is an addition number sentence. A number sentence with an equal sign is also called an **equation**.

To add numbers, first line up the digits by their place values. Then add. Remember that you sometimes need to regroup.

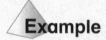 **Example**

Add: 426 + 398

Line up the digits in the ones, tens, and hundreds places.
Then add the ovnes: 6 ones + 8 ones = 14 ones

$$
\begin{array}{r}
1 \\
426 \\
+\ 398 \\
\hline
4
\end{array}
$$

← **Regroup 10 ones as 1 ten.**

Now, add the tens.
Remember to add the 1 ten that was regrouped.
1 ten + 2 tens + 9 tens = 12 tens

Regroup 10 tens as 1 hundred. →
$$
\begin{array}{r}
1\,1 \\
426 \\
+\ 398 \\
\hline
24
\end{array}
$$

Now, add the hundreds.
Remember to add the 1 hundred that was regrouped.
1 hundred + 4 hundreds + 3 hundreds = 8 hundreds

$$
\begin{array}{r}
1\,1 \\
426 \\
+\ 398 \\
\hline
824
\end{array}
$$

The sum is 824. (426 + 398 = 824)

Example

The Gomez family drove 379 miles on the first day of their vacation. On the second day, they drove 213 miles. How many miles did they drive during the two days?

Write an equation to help you solve the problem.

Let a ☐ stand for the unknown number.

$$379 + 213 = \boxed{}$$

Add 379 and 213.

```
    1    ← Regroup 10 ones as 1 ten.
  379
+ 213
─────
  592
```

The Gomez family drove 592 miles during two days.

Example

The PTA sold 267 orange juice boxes and 82 grape juice boxes at the All-School Track Meet. How many juice boxes did they sell in all?

Write an equation.

$$267 + 82 = \boxed{}$$

Add 267 and 82. Be careful lining up the digits.

```
   1    ← Regroup 10 tens as 1 hundred.
  267
+  82
─────
  349
```

Estimate to check the reasonableness of the answer.
    Round each number to its greatest place and add.
    267 rounds to 300. 82 rounds to 80.
    300 + 80 = 380
The estimated sum is 380. Because 380 is close to 349, the sum 349 is a reasonable answer.

The PTA sold 349 boxes of juice in all.

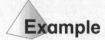

Example

On Monday, Devon practiced piano for 25 minutes and guitar for 30 minutes. On Monday, Corinne practiced piano for 45 minutes and violin for 15 minutes. Which student spent more time practicing music?

First, add to find the total number of minutes that Devon practiced. Write an equation. Let d (for Devon) equal the unknown number.
$25 + 30 = d$

Add: $25 + 30 = 55$
The sum of 25 and 30 is 55, so $d = 55$.

Use mental math to check the reasonableness of the answer.
    Add the tens: 2 tens + 3 tens = 5 tens
    Add the ones: 5 ones + 0 ones = 5 ones
    5 tens + 5 ones = 55
The mental math sum is 55, which matches the exact answer.

Devon practiced music for 55 minutes on Monday.

Now, add to find the total number of minutes that Corinne practiced. Write an equation. Let c (for Corinne) equal the unknown number.

$45 + 15 = c$

Add: $45 + 15 = 60$.
The sum of 45 and 15 is 60, so $c = 60$.

Use mental math to check the reasonableness of the answer.
    Add the tens: 4 tens + 1 ten = 5 tens
    Add the ones: 5 ones + 5 ones = 10 ones = 1 ten
    5 tens + 1 ten = 6 tens = 60
The mental math sum is 60, which matches the exact answer.

Corinne practiced music for a total of 60 minutes on Monday.

Compare the number of minutes each student practiced on Monday.

$55 < 60$

Corinne spent more time practicing music than Devon did.

CCSS: 3.OA.8, 3.NBT.2

When you add three numbers, it doesn't matter which two numbers you group together to add first. You can add numbers in any order. This is the associative property of addition. You learned it in Lesson 2.

Example

Vanessa has a stamp collection. She has 13 stamps from Europe, 18 U. S. stamps, and 7 stamps from Asia. How many stamps does Vanessa have in all?

Add: $13 + 18 + 7$

It might be easier to add 13 and 7 first. Use the commutative property to switch the order of the 7 and 18.

$13 + 7 + 18$

Use the associative property to group the first two addends.

$(13 + 7) + 18$

Add 13 and 7.

$13 + 7 = 20$

Now add 20 and 18.

$20 + 18 = 38$

The sum is 38.

Vanessa has 38 stamps in all.

Look back at the example above. If you add 13 and 18 first, and then add 7, you will still get 38.

$(13 + 18) + 7$

$31 \quad + 7 = 38$

You could also add 18 and 7 first, and then add 13 to get 38.

$13 + (18 + 7)$

$13 + \quad 25 \quad = 38$

Practice

Directions: For questions 1 through 12, find the sum.

1. $\begin{array}{r} 23 \\ + 15 \\ \hline \end{array}$

2. $\begin{array}{r} 19 \\ + 34 \\ \hline \end{array}$

3. $\begin{array}{r} 27 \\ + 55 \\ \hline \end{array}$

4. $\begin{array}{r} 120 \\ + 432 \\ \hline \end{array}$

5. $\begin{array}{r} 358 \\ + 167 \\ \hline \end{array}$

6. $\begin{array}{r} 501 \\ + 510 \\ \hline \end{array}$

7. $\begin{array}{r} 315 \\ + 173 \\ \hline \end{array}$

8. $\begin{array}{r} 294 \\ + 607 \\ \hline \end{array}$

9. $\begin{array}{r} 531 \\ + 352 \\ \hline \end{array}$

10. $\begin{array}{r} 580 \\ + 75 \\ \hline \end{array}$

11. $\begin{array}{r} 242 \\ + 92 \\ \hline \end{array}$

12. $\begin{array}{r} 717 \\ + 191 \\ \hline \end{array}$

CCSS: 3.OA.8, 3.NBT.2

Directions: For questions 13 through 18, set up the addition problem and find the sum.

13. 108 + 316 = _____

16. 625 + 48 = _____

14. 270 + 441 = _____

17. 260 + 91 = _____

15. 199 + 712 + 31 = _____

18. 83 + 176 + 304 = _____

Directions: For questions 19 through 21, write an equation. Then find the sum.

19. On Monday, Mika's family drove 114 miles from Los Angeles to Palm Springs. On Tuesday, they drove 289 miles from Palm Springs to Las Vegas. How far did Mika's family drive on Monday and Tuesday altogether?

_____

20. On Saturday, the snack bar sold 253 bags of popcorn. On Sunday, the snack bar sold 139 bags of popcorn. How many bags of popcorn were sold on Saturday and Sunday combined?

_____

21. The members of the PTA sold 308 raffle tickets before the Fall Festival. On the day of the Fall Festival, they sold 549 raffle tickets. How many raffle tickets were sold in all?

_____

22. On Thursday, Mr. Rios sold 187 pounds of tomatoes at his farm stand and 239 pounds of tomatoes at the downtown farmer's market. On Friday afternoon, he sold 268 pounds of tomatoes to local restaurants. How many pounds of tomatoes did Mr. Rios sell in all on Thursday and Friday?

_____

23. In June, Leon earned $166 mowing lawns and $45 helping a neighbor move. How much did Leon earn in all during June?

_____

24. Ms. Robbins spent $599 for a new computer. She also bought two new software programs. One program cost $207 and the other cost $144. How much did Ms. Robbins spend in all?

_____

25. Today's T-shirts Factory printed a radio station's logo on 412 short-sleeved T-shirts and 185 long-sleeved T-shirts. The factory also printed the same logo on 157 sweatshirts. How many logos did the factory print in all?

_____

26. On Monday morning, the librarian moved 147 books to a storage unit. On Wednesday, she sent 79 books to another library. How many books did the librarian remove from the library in all?

_____

CCSS: 3.OA.8, 3.NBT.2

Directions: Use the information in the table to answer questions 27 and 28.

The third-grade classes had a contest to see which class could collect the most pennies for an animal shelter. The table shows the number of pennies each class collected.

Pennies Collected by Students in the Third Grade

Class	Mrs. Johnson	Ms. Santiago	Mr. Potter
Pennies Collected	217	433	243

27. What was the total number of pennies collected by Mrs. Johnson's and Ms. Santiago's classes combined?

A. 640

B. 641

C. 650

D. 651

28. What was the total number of pennies collected by all the third-grade classes?

A. 833

B. 893

C. 894

D. 983

29. Ms. Ramirez wrote this addition sentence on the board.

$$48 + 52 + 27 = \boxed{}$$

What is the sum? _____

Explain how you used the associative property to find the sum.

_____

_____

_____

Lesson 6: Subtracting Whole Numbers

When you want to take away or find the number left over or find the difference between two numbers, you **subtract (–)**. The number you take away <u>from</u> is the **minuend**. The number you take away is the **subtrahend**. The answer when you subtract is the **difference**.

$$\text{minuend} \rightarrow \quad 17 \; - \; 8 \; = \; 9 \quad \leftarrow \text{difference}$$
$$\uparrow$$
$$\text{subtrahend}$$

$17 - 8 = 9$ is a subtraction number sentence. It is also an equation.

When you subtract numbers, first line up the digits according to their place values. Then subtract. Remember that you sometimes need to regroup.

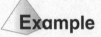 **Example**

       Subtract: 329 – 198

       Line up the digits in the ones, tens, and hundreds places.
       Then subtract the ones: 9 ones – 8 ones = 1 one

$$\begin{array}{r} 329 \\ -\,198 \\ \hline 1 \end{array}$$

       Now, move to the tens place. You cannot subtract 9 from 2, so you will
       need to regroup.

       Then subtract the tens: 12 tens – 9 tens = 3 tens

Regroup 1 hundred as 10 tens. $\rightarrow$
$$\begin{array}{r} \overset{2\;12}{3\!\!\!/29} \\ -\,198 \\ \hline 31 \end{array}$$

       Now, move to the hundreds place.
       Subtract: 2 hundreds – 1 hundred = 1 hundred

$$\begin{array}{r} \overset{2\;12}{3\!\!\!/29} \\ -\,198 \\ \hline 131 \end{array}$$

       The difference is 131. (329 – 198 = 131)

CCSS: 3.OA.8, 3.NBT.2

Example

A theatre had 782 tickets to sell for performances of a play. The theatre sold 254 tickets. How many tickets are left to sell?

Write an equation to help you solve the problem.

$782 - 254 = \boxed{}$

Subtract 254 from 782.

$$\begin{array}{r} {}^{7\,12} \\ 7\cancel{8}2 \\ -\ 254 \\ \hline 528 \end{array}$$ ← **Regroup 1 ten as 10 ones.**

There are 528 tickets left to sell.

Example

A store had 503 backpacks to sell. On the weekend before school starts, the store sold 379 backpacks. How many backpacks are left?

Write an equation.

$503 - 379 = \boxed{}$

Subtract 379 from 503.

Regroup 1 hundred as 10 tens. →
$$\begin{array}{r} {}^{9} \\ 4\,\cancel{10}13 \\ \cancel{5}\cancel{0}3 \\ -\ 379 \\ \hline 124 \end{array}$$ ← **Regroup 1 ten as 10 ones.**

Estimate to check that your answer is reasonable.
    Round each number to its greatest place and subtract.
    503 rounds to 500. 379 rounds to 400.
    $500 - 400 = 100$
The estimated difference is 100. Because 100 is close to 124, the answer 124 is reasonable.

There are 124 backpacks left.

Addition and subtraction are opposite operations. Subtraction "undoes" addition. Addition "undoes" subtraction. So, you can use addition to check your answer to a subtraction problem. When you add the difference and the number that you subtracted, your sum should be the number that you subtracted from. You can also use subtraction to check the answer to an addition problem.

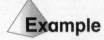

Example

On Friday morning, Tasty Bakery sold 632 muffins and 219 cookies. How many more muffins than cookies were sold?

Write an equation. Let *n* equal the difference between the number of muffins sold and the number of cookies sold.

$$632 - 219 = n$$

$$\begin{array}{r} {\scriptstyle 2\,12} \\ 6\cancel{3}2 \\ -\ 219 \\ \hline 413 \end{array}$$ ← **Regroup 1 ten as 10 ones.**

The difference between 632 and 219 is 413, so *n* = 413.

Use addition to check your answer.
Add the difference, 413, and the number subtracted, 219.

$$\begin{array}{r} {\scriptstyle 1} \\ 413 \\ +\ 219 \\ \hline 632 \end{array}$$ ← **Regroup 10 ones as 1 ten.**

The sum matches the number subtracted from, 632, so the difference 413 is correct.

On Friday morning, 413 more muffins were sold than cookies.

CCSS: 3.OA.8, 3.NBT.2

Sometimes you need to use both addition and subtraction to solve a problem.

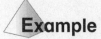 **Example**

Richie and Jana are on the swim team. Richie swam for 45 minutes on Monday and 65 minutes on Tuesday. Jana did not swim on Monday, but swam for 95 minutes on Tuesday. How many more minutes did Richie swim than Jana?

First, write an equation to find *r*, the total number of minutes that Richie swam.

$$45 + 65 = r$$

Add 45 and 65 to find the value of *r*.

$$\begin{array}{r} \overset{1}{4}5 \\ +\ 65 \\ \hline 110 \end{array}$$ ← **Regroup 10 ones as 1 ten.**

The sum of 45 and 65 is 110, so *r* = 110.

Richie swam for a total of 110 minutes.

Next, write an equation to find *d*, the difference between the number of minutes that Richie swam and the number of minutes that Jana swam.

$$110 - 95 = d$$

Subtract 95 from 110 to find the value of *d*.

$$\begin{array}{r} \overset{10}{\cancel{0}}\overset{}{0}10 \\ \cancel{1}\cancel{1}\cancel{0} \\ -\ 95 \\ \hline 15 \end{array}$$

The value of *d* is 15.

Use addition to check your answer.

Add the difference and the number subtracted.
$$15 + 95 = 110$$

The sum matches the number subtracted from, 110, so the difference 15 is correct.

Richie swam 15 minutes more than Jana.

Practice

Directions: For questions 1 through 12, find the difference.

1.
$$
\begin{array}{r}
54 \\
-\ 22 \\
\hline
\end{array}
$$

7.
$$
\begin{array}{r}
592 \\
-\ 167 \\
\hline
\end{array}
$$

2.
$$
\begin{array}{r}
63 \\
-\ 18 \\
\hline
\end{array}
$$

8.
$$
\begin{array}{r}
831 \\
-\ 359 \\
\hline
\end{array}
$$

3.
$$
\begin{array}{r}
809 \\
-\ 253 \\
\hline
\end{array}
$$

9.
$$
\begin{array}{r}
658 \\
-\ 77 \\
\hline
\end{array}
$$

4.
$$
\begin{array}{r}
41 \\
-\ 33 \\
\hline
\end{array}
$$

10.
$$
\begin{array}{r}
706 \\
-\ 415 \\
\hline
\end{array}
$$

5. $854 - 714 = $ _____

11. $811 - 37 = $ _____

6. $583 - 141 = $ _____

12. $700 - 136 = $ _____

13. Yesterday, the temperature in St. Louis, Missouri, was 76 degrees. Today, it was 69 degrees. How many degrees warmer was it yesterday?

_____

14. Jacob has 315 baseball cards in his collection. If he gives 120 of them to his brother, how many baseball cards will Jacob have left in his collection?

_____

15. Sally's Sandwich shop sold 59 tuna sandwiches, 127 ham sandwiches, and 196 chicken salads on Monday. Were more sandwiches or salads sold? How many more?

_____

16. A zoo has 162 birds in one walk-through display and 93 birds in another display. So far, the zookeeper has been able to check 118 of the birds this week. How many more birds need to be checked?

_____

17. Julie wants to attend a swim camp that costs $239 for one week. Her brother wants to go to a soccer camp that costs $575 for two weeks. How much more does two weeks of soccer camp cost than two weeks of swim camp?

_____

18. Treats for Pups sold 337 decorated dog dishes on their website last year. At the store in the mall, they sold 265 decorated dog dishes. How many more dishes were sold on the website than in the store?

   A. 52
   B. 62
   C. 72
   D. 102

19. A community theatre spent $597 on stage lights and $370 for a new curtain. How much more did it spend on the lights than on the curtain?

   A. $217
   B. $227
   C. $867
   D. $967

20. The cost to build raised garden beds for a community garden is $950. The members of the community garden club raised $537 in a plant sale. The club also won a grant of $200. How much more money do they need to build the raised beds?

   A. $213
   B. $413
   C. $737
   D. $750

21. A science museum director has $800 to buy a new telescope. The telescope costs $659. The shipping costs $50 extra. How much money will be left over if he buys the telescope?

   A. $91
   B. $141
   C. $151
   D. $709

22. Last year, the Spring Carnival raised $682. The PTA hoped to raise at least $250 more this year than last year. This year, the Spring Carnival raised $914. Did the PTA reach its goal this year?

   _____

Explain how you got your answer.

_____

_____

_____

Unit 1 Practice Test

1. Write the number four thousand seven hundred ninety in the following place-value table.

Thousands	Hundreds	Tens	Ones

How is the number from the place-value table written in standard form?

_____

2. Write the number three thousand eight hundred two in the following place-value table.

Thousands	Hundreds	Tens	Ones

How is the number from the place-value table written in expanded form?

_____

For questions 3 through 10, use the properties of addition to fill in the missing numbers. Then write the name of the property you used.

3. _____ + (5 + _____) = (6 + _____) + 4   _____

4. (_____ + 7) + 1 = 5 + (_____ + _____)   _____

5. _____ + 3 = 3 + 8   _____

6. 2 + _____ = 2   _____

7. 7 + 4 = _____ + 7   _____

8. (9 + _____) + 4 = _____ + (2 + _____)   _____

9. 5 + 6 = 6 + _____   _____

10. _____ + 0 = 15   _____

For questions 11 through 16, round each number to the given place.

11. Round **16** to the nearest ten. _____

12. Round **42** to the nearest ten. _____

13. Round **378** to the nearest hundred. _____

14. Round **265** to the nearest hundred. _____

15. Round **114** to the nearest ten. _____

16. Round **175** to the nearest ten. _____

Use the addition table to answer questions 17 and 18.

+	0	1	2	3	4	5
0	0	1	2	3	4	5
1	1	2	3	4	5	6
2	2	3	4	5	6	7
3	3	4	5	6	7	8
4	4	5	6	7	8	9
5	5	6	7	8	9	10

17. The numbers on either side of the diagonal line are the same. Which property best explains why?

_____

18. The numbers in the first unshaded row are the same as the numbers across the top of the addition table. Which property best explains why?

_____

For questions 19 through 24, find the sum or difference.

19. $\begin{array}{r} 43 \\ + 56 \\ \hline \end{array}$

20. $\begin{array}{r} 385 \\ + 347 \\ \hline \end{array}$

21. 593 + 384 = _____

22. $\begin{array}{r} 685 \\ - 447 \\ \hline \end{array}$

23. $\begin{array}{r} 703 \\ - 615 \\ \hline \end{array}$

24. 61 − 49 = _____

Use the following information to answer questions 25 through 27.

The third-grade classes had a contest to see which class could collect the most plastic water bottles for recycling. The following table shows the number of bottles each class collected.

Plastic Bottles Collected by Students in the Third Grade

Class	Mrs. Ortiz	Mr. Rocca	Ms. Jackson
Bottles Collected	228	674	252

25. What was the total number of bottles collected by Mrs. Ortiz's and Ms. Jackson's classes combined? _____

26. How many more bottles did Mr. Rocca's class collect than Ms. Jackson's class? _____

27. If Mrs. Ortiz's class collects 100 morve bottles, how many bottles would her class have? _____

Choose the correct answer.

28. Which shows an example of the commutative property?

    A. 8 + 0 = 8

    B. 6 + 6 = 10 + 2

    C. (3 + 4) = 5 = 3 + (4 + 5)

    D. 7 + 2 = 2 + 7

29. Latoya, Jane, and Andrew collected cans for recycling. Latoya collected 214 cans, Jane collected 232 cans, and Andrew collected 204 cans. How many cans did they collect in all?

    A. 640

    B. 648

    C. 650

    D. 750

30. How is 3208 written in expanded form?

    A. 300 + 20 + 8

    B. 3000 + 200 + 8

    C. 3000 + 200 + 80

    D. 3000 + 200 + 20 + 8

31. Kitchen Customs sold 651 aprons on their website last year. At the Kitchen Customs store in the mall, they sold 537 aprons. How many more aprons were sold on the website than in the store?

    A. 14

    B. 26

    C. 114

    D. 126

Use the following place-value table to answer questions 32 and 33.

Thousands	Hundreds	Tens	Ones
6	4	1	7

32. What is the value of the 7?

    A.   7

    B.   70

    C.   700

    D.   7000

33. What is the value of the 6?

    A.   6

    B.   60

    C.   600

    D.   6000

Solve each problem.

34. A play is performed on Friday night, Saturday afternoon, and Saturday night. For the Friday night show, 390 tickets were sold. For the shows on Saturday, 315 tickets and 259 tickets were sold. How many tickets were sold in all?

                _____

35. Zachary has 162 baseball cards and 47 football cards. Chris has 127 baseball cards and no football cards. How many sports cards do they have in all?

                _____

36. The library has 287 books about North America and 68 books about South America. There are 193 books about the rest of the world. How many more books are there about the Americas than about the rest of the world?

                _____

37. Mrs. Scott had $837 in a checking account. She wrote a check for $459 for a washing machine. She was given a rebate check for $25 when she bought the washing machine. She deposited the rebate check in her checking account. What is the balance in her checking account now?

                _____

38. In the wild, an elephant eats about 770 pounds of food each day. In a zoo, an elephant eats about 150 pounds of food each day. How much less food do two zoo elephants eat in one day than one wild elephant eats?

                _____

39. Happy Trails Adventures has two llamas that carry packs on hiking trips. The llamas are named Rosy and Woody. Rosy weighs 316 pounds and Woody weighs 392 pounds. Woody can carry a pack that weighs 85 pounds. Rosy can carry a pack that weighs 72 pounds.

Part A
What is Woody's total weight when he is carrying a full pack?

_____

Part B
Write an equation to find Rosy's total weight when she is carrying a full pack.

_____

Part C
Find how much more Woody weighs than Rosy when both are carrying a full pack.

_____

Part D
Is your answer to Part C reasonable? Explain your thinking.

_____

_____

_____

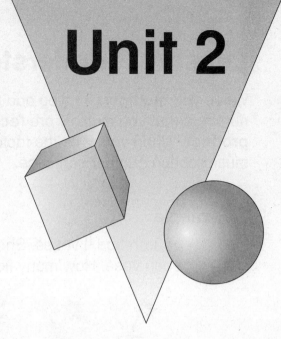

Unit 2

Operations and Algebraic Thinking

Multiplication and division can be used to solve many everyday problems. You can use multiplication to find the number of crayons in 3 boxes. You can use division to decide how to share a stack of crackers equally among 4 friends. Multiplication is also useful for finding how many tiles are needed to cover a rectangular floor. Store managers use multiplication to determine the number of cans of tomatoes in 6 cases or the cost of buying 5 cases of cereal. Doctors use division to determine how much medicine should be in each dose so that a patient receives the right amount throughout the day.

In this unit, you will represent and solve multiplication and division problems. You will also apply properties of numbers and identify multiplication patterns.

In This Unit

CCSS: 3.OA.1, 3.OA.3

Lesson 7: Understanding Multiplication

When you **multiply (×)**, you add the same number over and over again. The numbers that you multiply are **factors**. The answer when you multiply is the **product**. When you write the factors and the product together, you have a multiplication number sentence.

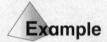

Example

Lauren had 4 vases. She bought some flowers. She put 2 flowers in each vase. How many flowers did she buy?

You can solve this problem using **skip counting**. Skip count by the number of flowers in each vase.

2, 4, 6, 8

You can also solve the problem using **repeated addition**. Add the number of flowers in each vase.

2 + 2 + 2 + 2 = 8

This is how you would solve the problem using multiplication.

4	×	2	=	8
↑		↑		↑
number of vases		number of flowers in each vase		total number of flowers
factor	×	factor	=	product

4 × 2 = 8 is a multiplication number sentence. It is also called an equation.

Lauren bought 8 flowers.

CCSS: 3.OA.1, 3.OA.3

A multiplication problem can also be solved using a **rectangular array**. A rectangular array is a group of objects arranged in the shape of a rectangle. The number of rows in the array is equal to one factor. The other factor is the number of objects in each row. The total number of objects in the array is the product.

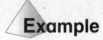 **Example**

Cassie arranged her collection of seashells in the following array.

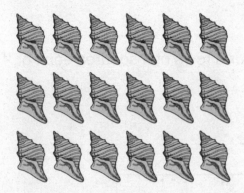

How many seashells does Cassie have?

You could count all the seashells. This may take too long if there are a lot of objects in an array. Instead, count the number of rows and the number of objects in each row. Then multiply the two numbers.

There are 3 rows. One factor is 3.

There are 6 seashells in each row. The other factor is 6.

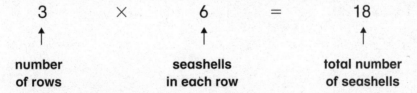

$$3 \qquad \times \qquad 6 \qquad = \qquad 18$$

| number of rows | seashells in each row | total number of seashells |

Cassie has 18 seashells.

Another way to solve a multiplication problem is by using an **area model**. Make a rectangle out of squares that are each 1 unit by 1 unit. The number of rows of squares in the rectangle equals one of the factors. The other factor is the number of squares in each row of the rectangle. The total number of squares that are in the rectangle is the product.

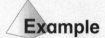 **Example**

Solve the following multiplication problem using an area model.

$6 \times 4 = \boxed{}$

Set up a rectangle that is either 6 rows of 4 squares each or 4 rows of 6 squares each.

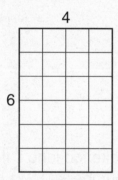 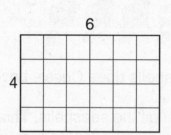

Count the total number of squares in each rectangle.
There are 24 squares in each rectangle.

Therefore, $6 \times 4 = 24$.

Also, $4 \times 6 = 24$.

CCSS: 3.OA.1, 3.OA.3

A multiplication problem can also be solved by using a number line. A number line helps you to see the order of numbers. You can solve a multiplication problem by making equal−sized jumps on a number line. The number of jumps you make is one of the factors. The number of tick marks you move in each jump is the other factor. The total number of tick marks you move in all of the jumps is the product.

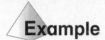

Example

Solve the following multiplication problem using a number line.

$5 \times 2 = \boxed{}$

Start at the zero mark on a number line. Make 5 jumps. Move 2 tick marks in each jump.

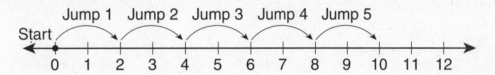

The number where you stop shows the total number of tick marks you have moved. This shows that the product is 10.

This problem can also be solved by making 2 jumps of 5 tick marks each.

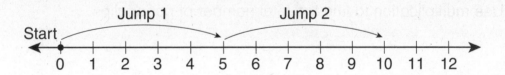

This way also shows that the product is 10.

Therefore, $5 \times 2 = 10$.

Also, $2 \times 5 = 10$.

Practice

Directions: Use the following array of paper clips to answer questions 1 through 3.

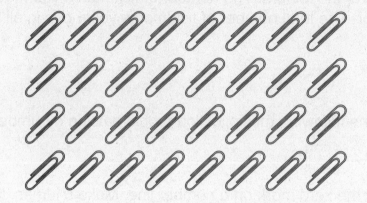

1. Use skip counting to find the total number of paper clips.

   _____ , _____ , _____ , _____

2. Use repeated addition to find the total number of paper clips.

   _____ + _____ + _____ + _____ = _____

3. Use multiplication to find the total number of paper clips.

   _____ × _____ = _____

Directions: Write multiplication equations to answer questions 4 through 7.

4. Katie put 4 cookies on each plate. There are 3 plates. How many cookies does Katie have?

   _____ × _____ = _____

CCSS: 3.OA.1, 3.OA.3

5. There are 4 peapods. There are 6 peas in each peapod. How many peas are there in all?

_____ × _____ = _____

6. What multiplication equation does the array of hammers show?

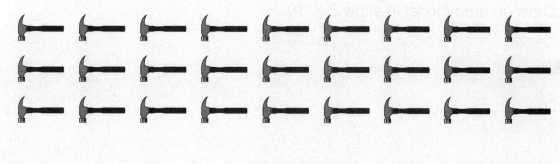

_____ × _____ = _____

7. What multiplication equation does the array of stars show?

_____ × _____ = _____

8. What multiplication equation does the following area model show?

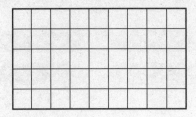

_____ ✕ _____ = _____

9. Draw an area model to show 3 ✕ 10.

3 ✕ 10 = _____

10. What multiplication equation is shown by the number line below?

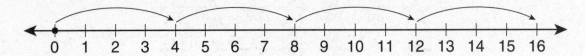

_____ ✕ _____ = _____

11. What multiplication equation is shown by the number line below?

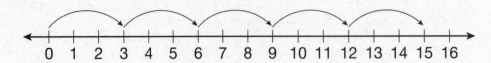

_____ ✕ _____ = _____

CCSS: 3.OA.1, 3.OA.3

12. Which multiplication equation does the following picture show?

    A. $3 \times 6 = 18$

    B. $3 \times 8 = 24$

    C. $9 \times 2 = 18$

    D. $3 \times 9 = 27$

13. Which multiplication equation does the following area model show?

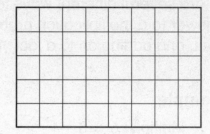

    A. $5 \times 8 = 40$

    B. $7 \times 5 = 35$

    C. $6 \times 8 = 48$

    D. $5 \times 7 = 35$

14. Use the following picture to answer the question.

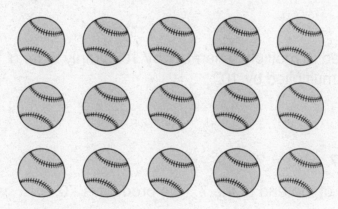

What is the total number of baseballs?

_____

Explain three different ways you could find the total number of baseballs.

_____

_____

_____

Lesson 8: Multiplication Facts

There are several different ways to use what you know about multiplication to find the answer to a multiplication problem. Repeated addition, skip counting, using doubles, and adding on to a fact you know are all good ways to find the answer.

Example

Multiply: $6 \times 3 = \square$

Think: 6 groups of 3.
Use repeated addition
to find the product.

The sum of the repeated addition
is 18, so the product is 18.

$6 \times 3 = 18$

$$
\begin{array}{r}
3 \\
3 \\
3 \\
3 \\
3 \\
+ 3 \\
\hline
18
\end{array}
$$

Whenever you need to multiply a number by 10, simply write a "0" at the end of the number being multiplied by 10.

Example

Multiply: $7 \times 10 = ?$

Write a 0 at the end of the 7. The product is 70.

$7 \times 10 = 70$

You can double a fact to help you find a product that you don't know.

Example

Multiply: $4 \times 6 = \square$

4 is a double of 2. You know that $2 \times 6 = 12$, so double the product of 2×6.

$12 + 12 = 24$, so the product of 4×6 is 24.

$4 \times 6 = 24$

CCSS: 3.OA.4, 3.OA.7

You can add on to a fact that you know to find a fact that you don't know.

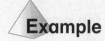**Example**

Multiply: $3 \times 8 = ?$

Start with the fact $2 \times 8 = 16$. Then add one more 8. $16 + 8 = 24$

$3 \times 8 = 24$

You can use a model and repeated addition to help find an unknown factor.

**Example**

What number makes the equation true?

$\boxed{} \times 7 = 21$

Use a model with rows of 7. The missing factor is the number of rows of 7 needed to make 21.

1 row of 7 = 7	7 + 7 = 14 2 rows of 7 = 14	7 + 7 + 7 = 21 3 rows of 7 = 21

Each row has 7. It takes 3 rows of 7 to have a total of 21.

$3 \times 7 = 21$

**Example**

What number makes the equation true?

$\boxed{} \times 5 = 45$

To find the missing factor, you can use a number line. Make jumps of 5.

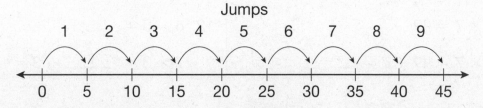

The number of jumps is the missing factor.

$9 \times 5 = 45$

Practice

Directions: For questions 1 through 4, use repeated addition to find the product.

1. 3×4

   $4 + 4 + 4 =$ _____

   $3 \times 4 =$ _____

2. 5×3

   $3 + 3 + 3 + 3 + 3 =$ _____

   $5 \times 3 =$ _____

3. 2×9

   $9 + 9 =$ _____

   $2 \times 9 =$ _____

4. 3×8

   $8 + 8 + 8 =$ _____

   $3 \times 8 =$ _____

Directions: For questions 5 through 8, find the first product. Then use doubling to find the second product.

5. $2 \times 7 =$ _____

   $4 \times 7 =$ _____

6. $3 \times 5 =$ _____

   $6 \times 5 =$ _____

7. $5 \times 4 =$ _____

   $5 \times 8 =$ _____

8. $7 \times 3 =$ _____

   $7 \times 6 =$ _____

Directions: For questions 9 through 12, find the first product. Then add on to find the second product.

9. $5 \times 9 = 45$

   $6 \times 9 = 45 + 9 =$ _____

10. $7 \times 7 = 49$

    $8 \times 7 = 49 + 7 =$ _____

11. $8 \times 8 =$ _____

    $9 \times 8 =$ _____ $+ 8 =$ _____

12. $6 \times 3 =$ _____

    $7 \times 3 =$ _____ $+ 3 =$ _____

Directions: For questions 13 through 16, find the product.

13. $8 \times 10 =$ _____

14. $5 \times 10 =$ _____

15. $10 \times 2 =$ _____

16. $10 \times 9 =$ _____

Directions: For questions 17 through 24, find the missing factor.

17. $3 \times \boxed{} = 18$ _____

18. $4 \times \boxed{} = 12$ _____

19. $\boxed{} \times 9 = 18$ _____

20. $\boxed{} \times 8 = 56$ _____

21. $6 \times \boxed{} = 60$ _____

22. $\boxed{} \times 7 = 49$ _____

23. $\boxed{} \times 6 = 48$ _____

24. $9 \times \boxed{} = 36$ _____

25. Which multiplication has a product of 27?

 A. 2×7

 B. 3×9

 C. 4×7

 D. 5×6

26. Which multiplication has a product of 35?

 A. 3×5

 B. 4×8

 C. 4×9

 D. 5×7

27. Darcy knows that $3 \times 3 = 9$. Explain how she could use that fact to find 6×3.

_____

_____

_____

What is the product of 6×3 ? _____

Lesson 9: Properties of Multiplication

The properties of multiplication can help you learn the multiplication facts more quickly and find other multiplication products mentally.

Commutative Property of Multiplication

This rule says that the **order** of the factors **does not change the product.**

$$3 \times 2 = 2 \times 3$$
$$6 = 6$$

Look at the arrays of stars. Switching the number of rows with the number of columns gives the same number of stars in the array.

| 3 rows, 2 columns | 2 rows, 3 columns |
| 6 stars in all | 6 stars in all |

Example

What number can replace the ☐ to make the equation true?

$$5 \times 7 = \boxed{} \times 5$$

According to the commutative property, the order of the numbers you are multiplying does not change the product. So, 5×7 is equal to 7×5.

The number 7 makes the equation true. $5 \times 7 = 7 \times 5$

Associative Property of Multiplication

This rule says that when you multiply three numbers, the way in which you **group** the numbers **does not change the product**. Parentheses ( ) show which two numbers you multiply first.

$$(5 \times 2) \times 4 = 5 \times (2 \times 4)$$
$$10 \times 4 = 5 \times 8$$
$$40 = 40$$

CCSS: 3.OA.5

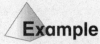 **Example**

Find the missing number in the equation below.

$(2 \times 4) \times 3 = 2 \times (4 \times \boxed{})$

According to the associative property, the grouping of the numbers does not change the product. The number sentence is $2 \times 4 \times 3$ no matter which way the numbers are grouped.

The missing number is 3. $(2 \times 4) \times 3 = 2 \times (4 \times 3)$

Identity Property of Multiplication

This rule says that when you **multiply 1 and any number**, the product is **that same number**.

$7 \times 1 = 7$ $1 \times 4 = 4$

Example

What number can replace the $\boxed{}$ to make the equation true?

$9 \times \boxed{} = 9$

According to the identity property, the product of any number and 1 is that number. 1 is the only number that can be multiplied by 9 to get a product of 9.

The number 1 makes the equation true. $9 \times 1 = 9$

Distributive Property

This rule says that multiplying the sum of two numbers by a third number is the same as multiplying each addend by the third number and adding the products.

$$2 \times (3 + 4) = (2 \times 3) + (2 \times 4)$$
$$2 \times 7 = 6 + 8$$
$$14 = 14$$

This means that you can use the distributive property to separate a multiplication problem into the sum of two easier problems.

The model below shows one way to use the distributive property to find the product of 6 × 7. The first array has 6 rows with 7 shapes in each row. To use the distributive property, separate the array into two smaller arrays. The second array shows 6 rows of 5 shapes and 6 rows of 2 shapes.

6 × 7	6 × 5	+	6 × 2
42	30	+	12
		42	

The model shows that:

$$6 \times 7 = 6 \times (5 + 2)$$
$$= (6 \times 5) + (6 \times 2)$$
$$= 30 + 12$$
$$= 42$$

The distributive property shows why the strategies of adding on and doubling can help you find products. You can separate a fact you don't remember into two facts that you do remember.

CCSS: 3.OA.5

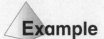 **Example**

Multiply: 8 × 6

Separate the problem into two easier problems.

$$8 \times 6 = 8 \times (5 + 1)$$
$$= (8 \times 5) + (8 \times 1)$$
$$= \quad 40 \quad + \quad 8$$
$$= \quad\quad\quad 48$$

8 × 6 = 48

You can use the properties of multiplication to multiply three numbers mentally.

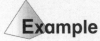 **Example**

Multiply: 3 × 2 × 4

Use the associative property to group the first two factors.

(3 × 2) × 4

Multiply the grouped factors: 3 × 2 = 6

$$\underbrace{(3 \times 2)}_{} \times 4$$
$$6 \times 4 = 24$$

3 × 2 × 4 = 24

 **Example**

Multiply: 2 × 3 × 5

Multiplying by 10 is easy to do mentally. Use the commutative property to change the order of the factors 3 and 5.

$$2 \times 3 \times 5 = 2 \times 5 \times 3$$

Use the associative property to group the first two factors.

$$\underbrace{(2 \times 5)}_{} \times 3$$
$$10 \times 3 = 30$$

2 × 3 × 5 = 30

65

Practice

Directions: For questions 1 through 4, use the drawings and the multiplication properties to help you complete the multiplication sentences.

1.

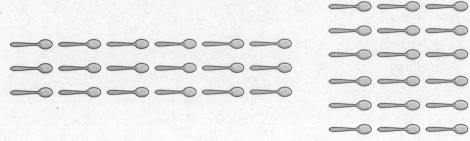

_____ × 6 = _____ × _____

2.

5 × _____ = (_____ × 5) + (_____ × 4)

3.

(5 × 2) × _____ = _____ × (2 × 3)

4.

$1 \times \underline{\hspace{1cm}} = 5$

Directions: For questions 5 through 12, use the properties of multiplication to fill in the missing numbers.

5. $10 \times \underline{\hspace{1cm}} = 3 \times 10$

6. $\underline{\hspace{1cm}} \times 1 = 7$

7. $(4 \times \underline{\hspace{1cm}}) \times 6 = 4 \times (2 \times 6)$

8. $\underline{\hspace{1cm}} \times 5 = 5 \times 9$

9. $1 \times \underline{\hspace{1cm}} = 2$

10. $3 \times 9 = (3 \times \underline{\hspace{1cm}}) + (3 \times 4)$

11. $\underline{\hspace{1cm}} \times (6 + 3) = (5 \times 6) + (5 \times 3)$

12. $5 \times (11 \times \underline{\hspace{1cm}}) = (\underline{\hspace{1cm}} \times \underline{\hspace{1cm}}) \times 2$

Directions: For questions 13 through 16, use the properties of multiplication to fill in the missing numbers. Then write the name of the property you used.

13. $4 \times 5 = 5 \times \underline{\hspace{1cm}}$ _____

14. $5 \times (3 \times 2) = (5 \times 3) \times \underline{\hspace{1cm}}$ _____

15. $7 \times 5 = (7 \times 2) + (7 \times \underline{\hspace{1cm}})$ _____

16. $1 \times \underline{\hspace{1cm}} = 6$ _____

17. Draw a diagram to show that $2 \times 5 = 5 \times 2$.

18. Draw a diagram to show that $1 \times 8 = 8$.

19. Multiply the following three numbers. Show your work.

    $2 \times 3 \times 3$

20. Multiply the three numbers from question 19 in a different order. Show your work.

    Is the answer the same as in question 19? _____

21. Samantha wants to use mental math to solve this problem: $7 \times 5 \times 2$. How could she solve this problem mentally?

    _____

    _____

    _____

22. Which shows an example of the identity property of multiplication?

    A. $2 \times 6 = 12$

    B. $6 \times 1 = 6$

    C. $2 \times 3 = 3 \times 2$

    D. $(2 \times 6) \times 3 = 2 \times (6 \times 3)$

23. Tracy has 1 page of math homework. There are 9 problems on the page. How many math problems does Tracy have for homework?

    A. 0

    B. 1

    C. 9

    D. 18

24. Which shows using the distributive property to find the product of 8 and 9?

    A. $(4 \times 4) + (4 \times 5)$

    B. $(4 \times 9) + (4 \times 9)$

    C. $(8 \times 5) + (8 \times 5)$

    D. $(8 \times 4) + (8 \times 5)$

25. Randy and Farah have the same number of stamps. Randy puts all his stamps in 6 rows with 4 stamps in each row. Farah puts all her stamps in 4 rows. How many stamps does Farah put in each row?

    A. 4

    B. 6

    C. 12

    D. 24

26. Mari is playing a number guessing game. Mari says "4 times two other numbers is 40." What are Mari's numbers? Hint: 1 is not one of her numbers.

    _____

    Explain how you found your answer.

    _____

    _____

    _____

CCSS: 3.OA.9

Lesson 10: Multiplication Patterns

When you look at a multiplication table, you can see patterns in the rows and columns. These patterns make it easier to learn and remember the multiplication facts.

×	1	2	3	4	5	6	7	8	9	10
1	1	2	3	4	5	6	7	8	9	10
2	2	4	6	8	10	12	14	16	18	20
3	3	6	9	12	15	18	21	24	27	30
4	4	8	12	16	20	24	28	32	36	40
5	5	10	15	20	25	30	35	40	45	50
6	6	12	18	24	30	36	42	48	54	60
7	7	14	21	28	35	42	49	56	63	70
8	8	16	24	32	40	48	56	64	72	80
9	9	18	27	36	45	54	63	72	81	90
10	10	20	30	40	50	60	70	80	90	100

A **multiple** of a whole number is found by multiplying that number by any other whole number.

For example, 6 is a multiple of 2 because $2 \times 3 = 6$.

Example

In the multiplication table, what pattern do you notice in the multiples of 2?

Look at the row for the factor 2.

2	2	4	6	8	10	12	14	16	18	20

All the multiples of 2 are even numbers.

CCSS: 3.OA.9

Example

If you draw a diagonal from the upper left to the lower right of the multiplication table, the numbers on either side of the diagonal match. Which property of multiplication explains this pattern?

Pick two numbers: for example, 3 and 4. The table shows that there is a matching product of 12 on either side of the diagonal whether you multiply 3×4 or 4×3.

×	1	2	3	4	5
1	1	2	3	4	5
2	2	4	6	8	10
3	3	6	9	12	15
4	4	8	⑫	6	20
5	5	10	15	20	25

×	1	2	3	4	5
1	1	2	3	4	5
2	2	4	6	8	10
3	3	6	9	⑫	15
4	4	8	12	6	20
5	5	10	15	20	25

Think of the properties of multiplication. The commutative property says that if you change the order of the numbers to be multiplied, the product stays the same.

The commutative property explains why the numbers on either side of the diagonal are the same.

Even and Odd Products

When you multiply numbers, you can know if the product will be even or odd.

If you multiply two even numbers, the product will be even. $2 \times 4 = 8$ $6 \times 8 = 48$

If you multiply two odd numbers, the product will be odd. $7 \times 5 = 35$ $9 \times 1 = 9$

If you multiply an even number and an odd number, the product will be even. $2 \times 5 = 10$ $8 \times 3 = 24$

Practice

Directions: For questions 1 through 3, use the part of the multiplication table shown below.

9	9	18	27	36	45	54	63	72	81	90

1. Find the sum of the digits for each multiple of 9.

   The sum of the digits for 9 is 9.

   The sum of the digits for 18 is 1 + 8 = 9.

   The sum of the digits for 27 is ____ + ____ = ____.

   The sum of the digits for 36 is ____ + ____ = ____.

   The sum of the digits for 45 is ____ + ____ = ____.

   The sum of the digits for 54 is ____ + ____ = ____.

   The sum of the digits for 63 is ____ + ____ = ____.

   The sum of the digits for 72 is ____ + ____ = ____.

   The sum of the digits for 81 is ____ + ____ = ____.

   The sum of the digits for 90 is ____ + ____ = ____.

2. What do you notice about the sum of the digits for each multiple of 9?

   _____

3. In question 1, look at the multiples of 9 from 18 to 90. What pattern do you notice in the multiples of 9?

   _____

CCSS: 3.OA.9

4. In the multiplication table, what pattern do you notice in the multiples of 5?

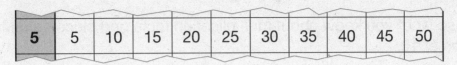

| 5 | 5 | 10 | 15 | 20 | 25 | 30 | 35 | 40 | 45 | 50 |

_____

_____

5. In the multiplication table, what pattern do you notice in the multiples of 10?

| 10 | 10 | 20 | 30 | 40 | 50 | 60 | 70 | 80 | 90 | 100 |

_____

_____

6. Devon multiplied a number by 7. The product was odd. Which could be the other factor that Devon multiplied?

A. 4

B. 6

C. 8

D. 9

7. Which property best explains why the row for 6 and the column for 6 in the multiplication table have the same numbers?

A. associative property of multiplication

B. commutative property of multiplication

C. distributive property

D. identity property of multiplication

8. Paige says that when you multiply an even number by an odd number, the product is even. Gilberto disagrees. He says that the product is an odd number. Who is correct?

_____

Give 3 examples to support your answer.

Lesson 11: Multiplying by Multiples of 10

You can use what you know about multiplication facts and place value to multiply by multiples of 10.

The multiples of ten are 10, 20, 30, 40, 50, and so on. You can think of these numbers as 1 ten, 2 tens, 3 tens, and so on.

Example

Multiply: 3×20

Think of 20 as 2 tens and use the multiplication fact $3 \times 2 = 6$.

$3 \times 20 = \underbrace{3 \times 2}_{\text{6 tens}} \text{ tens}$

6 tens = 60

$3 \times 20 = 60$

Think about what you know about place value:

8 tens is 80.
9 tens is 90.

As soon as you have more than 9 tens, you can use the hundreds place.

10 tens is 100.
12 tens is 120.

You can think of the place-value name *tens* as taking the place of a zero in the ones place.

Example

Multiply: 7×50

Think of 50 as 5 tens and use the multiplication fact $7 \times 5 = 35$.

$7 \times 50 = \underbrace{7 \times 5}_{\text{35 tens}} \text{ tens}$

35 tens = 350

$7 \times 50 = 350$

CCSS: 3.NBT.3

Sometimes a multiplication fact gives a product that is a multiple of ten, such as 20 or 30. Keep careful track of the zeros in the product.

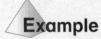

Example

Multiply: 5×60

Think of 60 as 6 tens, and use the multiplication fact $5 \times 6 = 30$. Notice that this fact has a zero in the ones place of the product.

$$5 \times 60 = \underbrace{5 \times 6}_{\text{30 tens}} \text{ tens}$$

30 tens = 300

$5 \times 60 = 300$

Practice

Directions: For questions 1 through 6, complete the multiplication fact you would use and write the product.

1. 3×40

    Use the fact:

    $3 \times 4 = $ _____

    $3 \times 40 = $ _____

2. 4×20

    Use the fact:

    _____ $\times$ _____ = _____

    $4 \times 20 = $ _____

3. 7×30

    Use the fact:

    _____ $\times$ _____ = _____

    $7 \times 30 = $ _____

4. 50×9

    Use the fact:

    _____ $\times$ _____ = _____

    $50 \times 9 = $ _____

5. 60×8

    Use the fact:

    _____ $\times$ _____ = _____

    $60 \times 8 = $ _____

6. 5×40

    Use the fact:

    _____ $\times$ _____ = _____

    $5 \times 40 = $ _____

Directions: For questions 7 through 18, write the product.

7. 8 × 80 = _____

8. 30 × 6 = _____

9. 7 × 60 = _____

10. 8 × 90 = _____

11. 70 × 4 = _____

12. 6 × 50 = _____

13. 4 × 80 = _____

14. 9 × 20 = _____

15. 90 × 6 = _____

16. 7 × 70 = _____

17. 9 × 90 = _____

18. 80 × 2 = _____

19. Which has a product of 360?

    A. 4 × 90

    B. 5 × 70

    C. 6 × 50

    D. 9 × 30

20. Multiply: 50 × 8

    A. 40

    B. 45

    C. 400

    D. 500

21. Tyrone says that when he multiplies 2 × 40, he thinks 2 × 4 × 10. Do you think he will get the correct answer?

        _____

Explain your answer.

_____

_____

_____

CCSS: 3.OA.3, 3.OA.8

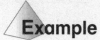

Lesson 12: Solving Problems with Multiplication

You can use multiplication facts to solve word problems.

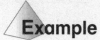 **Example**

Sara has 4 baskets. She puts 3 eggs in each basket. How many eggs does Sara have?

Write an equation to help you find the total number of eggs.
Let a ⬚ stand for the unknown number.

4 baskets × 3 eggs in each basket = total number of eggs

$4 \times 3 = $ ⬚

Use repeated addition to find the unknown number.
   $3 + 3 + 3 + 3 = 12$
$4 \times 3 = 12$

Sara has 12 eggs.

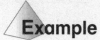 **Example**

Mrs. Turner wants to make 6 bows. For each bow, she needs 8 inches of ribbon. How many inches of ribbon does Mrs. Turner need in all?

Write an equation to help you find how much ribbon Mrs. Turner needs.
6 bows × 8 inches of ribbon for each bow = total inches of ribbon needed

$6 \times 8 = $ ⬚

Double a fact to find the unknown number.
   6 is a double of 3, and $3 \times 8 = 24$.
   Double the product of 3×8 to find the product of 6×8.
   $24 + 24 = 48$
$6 \times 8 = 48$

Mrs. Turner needs 48 inches of ribbon.

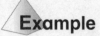**Example**

In Mr. Foster's classroom, there are 4 rows of desks. There are 5 desks in each row. How many desks are in Mr. Foster's classroom?

Write an equation to help you find the number of desks in Mr. Foster's classroom.

4 rows × 5 desks in each row = total number of desks

$4 \times 5 = \boxed{}$

Use skip-counting to find the unknown number.
   Skip count by 5s, the number of desks in each row.
   5, 10, 15, 20

$4 \times 5 = 20$

There are 20 desks in Mr. Foster's classroom.

Example

A book costs $6. A toy robot costs 5 times as much as the book. How much does the toy robot cost?

Write an equation to help you find how much the toy robot costs.
5 times $6 (the cost of the book) = the cost of the toy robot

$5 \times 6 = \boxed{}$

Use repeated addition to find the unknown number.
   $6 + 6 + 6 + 6 + 6 = 30$
$5 \times 6 = 30$

The toy robot costs $30.

 **Example**

A table is 30 inches wide. The table is 2 times longer than it is wide. How long is the table?

Write an equation to help you find the length of the table.
2 times 30 inches = the length of the table in inches

$2 \times 30 = \boxed{}$

Think of 30 as 3 tens. Use the multiplication fact $2 \times 3 = 6$.
    $2 \times 30 = 2 \times 3$ tens $= 6$ tens
    6 tens $= 60$
$2 \times 30 = 60$

The table is 60 inches long.

Example

Mason sells fruit baskets at his store. A small basket contains 5 apples and 4 oranges. What is the total number of pieces of fruit in 10 small baskets?

First, add to find the total amount of fruit in each basket.
Write an equation. Let *f* equal the total number of pieces of fruit in each basket.
$5 + 4 = f$

Add: $5 + 4 = 9$, so $f = 9$.

Each basket contains 9 pieces of fruit.

Then, multiply to find the total amount of fruit in 10 baskets.
Write an equation. Let *n* equal the unknown number of pieces of fruit.
$10 \times 9 = n$

Remember, whenever you multiply 10 and another number, write a 0 at the end of the other number.
To find the product of 10×9, write a 0 at the end of the 9.
$10 \times 9 = 90$

There are 90 pieces of fruit in 10 small fruit baskets.

Practice

1. Yancey bought a box of rock samples at a natural history museum. The rock samples were arranged in the box in 3 rows and 6 columns. How many rock samples were there?

    number of rows × number of columns = number of rock samples

    _____   ×   _____   =   _____

    There were _____ rock samples.

2. Nick bought 9 hamburgers. Each hamburger cost $2. How much did Nick spend on hamburgers?

    number of hamburgers × cost of each = amount spent on hamburgers

    _____   ×   _____   =   _____

    Nick spent _____ on hamburgers.

Directions: For questions 3 through 5, write an equation. Then solve the equation.

3. Chan bought 4 packages of sports cards. Each package had 8 cards. How many cards did Chan buy in all?

    _____

4. Ms. Hom bought 5 cans of peanuts. Each can of peanuts cost $3. How much did she spend on the peanuts?

    _____

5. Cass has 2 containers of tomato plants. Each container has 6 tomato plants. How many tomato plants does Cass have in all?

    _____

CCSS: 3.OA.3, 3.OA.8

6. There are 5 vases. Each vase has 7 flowers. How many flowers are there in all?

    _____

7. Jess has 6 model cars on each shelf. He has 6 shelves. How many model cars does Jess have in all?

    _____

8. There are 9 baskets of peaches. Each basket holds 7 peaches. How many peaches are there in all?

    _____

9. Mrs. Evans bought a sheet of stamps. The sheet of stamps had 10 rows and 4 columns. How many stamps are there in all?

    _____

10. Jordan has a rope that is 5 feet long. Marshall has a rope that is 4 times as long as Jordan's rope. How long is Marshall's rope?

    _____

11. A sandwich costs $5 and a drink costs $2. A pizza costs 2 times as much as a sandwich and a drink combined. What is the cost of the pizza?

    _____

12. Andrew paid $3 each for 6 books. David paid $4 each for 5 books. Who spent more for his books?

    _____

13. For a class play, there are 5 rows of chairs. Each row has 9 chairs. How many chairs are there in all?

    A. 36

    B. 40

    C. 45

    D. 50

14. Greta has 7 rows of potatoes. Each row has 7 potato plants. How many potato plants does Greta have in all?

    A. 49

    B. 48

    C. 42

    D. 14

15. Tickets to a play cost $4 for students and twice as much for adults. What is the total cost of tickets for 3 adults?

    A. $8

    B. $12

    C. $20

    D. $24

16. Mika arranged a display of pumpkins. She made 4 rows of 9 pumpkins. How many pumpkins were in the display?

    A. 5

    B. 13

    C. 36

    D. 49

17. Maggie buys 5 pounds of oranges each week for her family. She buys 4 times as many pounds of oranges each week for her café. How many pounds of oranges does Maggie buy each week in all?

    _____

Explain how you found your answer.

_____

_____

_____

CCSS: 3.OA.2, 3.OA.3

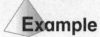

Lesson 13: Understanding Division

When you **divide (÷)** a number or a set of things, you break the total amount into equal groups. The number you are dividing is the **dividend**. The number you are dividing by is the **divisor**. The answer when you divide is the **quotient**.

Example

Mr. Morrison gave a total of 12 gold stars to some students. He gave each student 4 gold stars. How many students received gold stars?

You can solve this problem by separating the gold stars into groups. Each student received 4 gold stars, so put the gold stars into groups of 4.

There are 3 groups of 4 gold stars.

You can also solve the problem using **repeated subtraction**. Start with the total number of gold stars that were given to the students. Then subtract the number of gold stars that each student received. Repeat the subtraction until you cannot subtract any more. Count the number of times you subtracted.

$12 - 4 = 8$ (1 time)

$8 - 4 = 4$ (2 times)

$4 - 4 = 0$ (3 times)

You subtracted 4 from 12 a total of 3 times.

This is how you would solve the problem using division.

$$12 \quad ÷ \quad 4 \quad = \quad 3$$

↑	↑	↑
total number of gold stars	**number of gold stars for each student**	**number of students**
dividend ÷	**divisor** =	**quotient**

Mr. Morrison gave the 12 gold stars to 3 students.

A division problem can be solved using **equal sharing**. You use equal sharing when you know how many groups there are and you want to find out how many objects there are in each group. Each group will have an equal share of the whole amount. Start with the total number of objects and place one object in each group. Then start again and place another object in each group. Keep going until all the objects have been placed in a group. The number of objects in each group is the quotient.

Example

Kim and Maria combined their money to buy a pack of gum. There are 10 sticks of gum in a pack. They want to divide the gum equally between themselves. How many sticks of gum will each girl get?

Label 1 stick of gum as Kim's and 1 as Maria's. Repeat until there are no sticks of gum left.

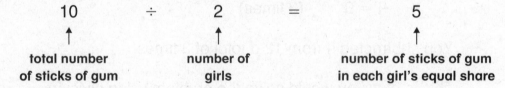

Kim   Maria   Kim   Maria   Kim   Maria   Kim   Maria   Kim   Maria

Count the number of sticks of gum labeled for each girl.

There are 5 sticks of gum for each girl.

This is how you would solve the problem using division.

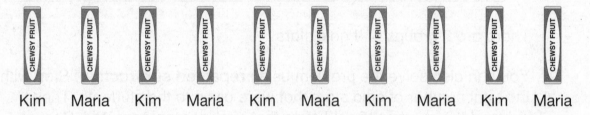

$$10 \div 2 = 5$$

total number of sticks of gum   number of girls   number of sticks of gum in each girl's equal share

The dividend is 10. The divisor is 2. The quotient is 5.

$10 \div 2 = 5$ is a division number sentence. It is also called an equation.

Each girl will get 5 sticks of gum.

CCSS: 3.OA.2, 3.OA.3

The same type of **rectangular array** that can be used to solve a multiplication problem can be used to solve a division problem. The total number of objects in the dividend is made into an array. Each row of the array has the same number of objects as the divisor. The quotient will equal the number of rows in the array.

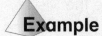

Example

Steve has 15 collectible stamps. He wants to divide them equally among 5 of his friends. How many stamps will each friend receive?

Steve made the following array of the 15 stamps. Notice that there are 5 stamps in each row.

There are 3 rows of stamps.

This is how you would solve the problem using division.

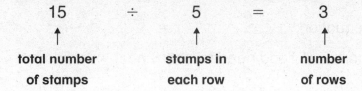

$$15 \div 5 = 3$$

total number of stamps   stamps in each row   number of rows

The dividend is 15. The divisor is 5. The quotient is 3.

Each friend will receive 3 stamps.

Note that the numbers in the Example above represent two different things: stamps and friends. This happens a lot in multiplication and division problems. You are dealing with two kinds of objects. This is different than addition and subtraction problems, where you are always dealing with one kind of object. For example, a subtraction problem might ask you how many apples you would have if you started with 10 and gave 5 to a friend. In that case, you are dealing with only one kind of object: apples.

Practice

1. There are 24 students in Mr. Cody's class. There are 6 students sitting at each table in the classroom. How many tables are there in Mr. Cody's classroom? Use repeated subtraction in the space below to solve the problem.

   Write a division equation: _____ ÷ _____ = _____

   There are _____ tables in Mr. Cody's classroom.

2. Todd is taking his friends golfing at Valley View golf course. There are 8 people going, and 2 people can ride in each golf cart.

   Write a division equation: _____ ÷ _____ = _____

   How many golf carts will Todd need for the 8 people to ride in? _____

3. At a party, 27 balloons are being given out. Each child at the party gets 3 balloons. How many children are at the party? Use repeated subtraction to solve the problem.

   Write a division equation: _____ ÷ _____ = _____

   There are _____ children at the party.

CCSS: 3.OA.2, 3.OA.3

4. Josh, Mya, and Eli helped their grandpa rake leaves. He gave them $12 to share equally among the 3 of them. How much money will each grandchild get? Use equal sharing by writing the name of a grandchild under each of the following dollar bills. Then count the number of dollar bills for each grandchild.

_____ _____ _____ _____ _____ _____

_____ _____ _____ _____ _____

Write a division equation: _____ ÷ _____ = _____

Each grandchild will get $_____.

5. A greeting card company sells birthday cards in packs of 5. How many packs can the company make if it has the following 30 birthday cards? Use grouping to solve the problem.

Write a division equation: _____ ÷ _____ = _____

The company can make _____ packs of birthday cards.

Directions: Use the following information to answer questions 6 and 7.

Gina has 12 ripe bananas. She wants to use the bananas to make banana bread. She needs 2 bananas to make each loaf of bread.

6. How many loaves of banana bread can Gina make?

   A. 3

   B. 4

   C. 6

   D. 9

7. Which equation can be used to solve this problem?

   A. $12 \div 2 = 6$

   B. $12 \div 3 = 4$

   C. $12 \div 4 = 3$

   D. $14 \div 2 = 7$

8. Jillian put a new tile floor in her kitchen. She used 42 tiles. The tiles are in 6 equal rows. How many tiles are in each row?

   Use a drawing to solve the problem.

   There are _____ tiles in each row.

   Explain how you found your answer.

   _____

   _____

CCSS: 3.OA.4, 3.OA.6, 3.OA.7

Lesson 14: Division Facts

There are different ways to use what you know about multiplication and division to help you find the answer to a division problem. If you know the multiplication facts, you can use them to help you remember the division facts.

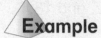 **Example**

What is 28 ÷ 7?

You can use the multiplication table to show the division facts. Look across the top of the table for 7. Go down in the 7 column until you get to 28. Then go across to the number on the left. The quotient is 4.

×	1	2	3	4	5	6	7	8	9	10
1	1	2	3	4	5	6	7	8	9	10
2	2	4	6	8	10	12	14	16	18	20
3	3	6	9	12	15	18	21	24	27	30
4	4	8	12	16	20	24	(28)	32	36	40
5	5	10	15	20	25	30	35	40	45	50

So, 28 ÷ 7 = 4.

You can also use repeated subtraction to help you find a quotient.

 Example

Divide: 40 ÷ 8 = ☐

Use repeated subtraction to find the quotient

40 − 8 = 32	1 time
32 − 8 = 24	2 times
24 − 8 = 16	3 times
16 − 8 = 8	4 times
8 − 8 = 0	5 times

You subtracted 8 from 40 a total of 5 times, so the quotient is 5.

40 ÷ 8 = 5

CCSS: 3.OA.4, 3.OA.6, 3.OA.7

Multiplication and division are "opposites." You might say that multiplication and division "undo" each other. How does this work? Multiply any two numbers. Then take this product and divide it by either of the two numbers. The quotient will be the other number. Fact families show how three numbers are related by multiplication and division.

Here is the multiplication and division fact family for the numbers 9, 2, and 18.

$9 \times 2 = 18$ $18 \div 2 = 9$
$2 \times 9 = 18$ $18 \div 9 = 2$

Example

Divide: $24 \div 6 = ?$

Think of a related multiplication fact.
$? \times 6 = 24$

How many 6s equal 24?

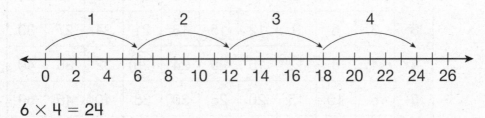

$6 \times 4 = 24$

So, $24 \div 6 = 4$.

Example

What number makes the equation true?

$72 \div \boxed{} = 9$

What other number is in the fact family with 9 and 72?
Divide 72 by 9 to find out.
$72 \div 9 = 8$

So, $72 \div 8 = 9$.

CCSS: 3.OA.4, 3.OA.6, 3.OA.7

Example

What number makes the equation true?

$\boxed{} \div 9 = 10$

Think of a related multiplication fact.

$9 \times 10 = \boxed{}$

$9 \times 10 = 90$

So, $90 \div 9 = 10$.

Practice

Directions: For questions 1 through 10, find the quotient.

1. $63 \div 7 = $ _____

2. $32 \div 8 = $ _____

3. $15 \div 5 = $ _____

4. $81 \div 9 = $ _____

5. $18 \div 6 = $ _____

6. $49 \div 7 = $ _____

7. $54 \div 9 = $ _____

8. $24 \div 8 = $ _____

9. $21 \div 3 = $ _____

10. $56 \div 7 = $ _____

Directions: For questions 11 through 16, find the missing number.

11. $8 \div \boxed{} = 2$ _____

12. $\boxed{} \div 8 = 7$ _____

13. $36 \div \boxed{} = 6$ _____

14. $\boxed{} \div 4 = 5$ _____

15. $\boxed{} \div 2 = 10$ _____

16. $35 \div \boxed{} = 7$ _____

17. Which division has a quotient of 6?

   A. 18 ÷ 3

   B. 30 ÷ 4

   C. 40 ÷ 8

   D. 63 ÷ 7

18. Which division has a quotient of 8?

   A. 14 ÷ 2

   B. 24 ÷ 4

   C. 54 ÷ 6

   D. 64 ÷ 8

Directions: For questions 19 and 20, complete the fact families.

19. multiplication equations

   $6 \times 7 = 42$

   _____

   division equations

   _____

   _____

20. multiplication equations

   _____

   _____

   division equations

   _____

   $32 \div 4 = 8$

21. Write the multiplication and division fact family for 5, 6, and 30.

   _____   _____

   _____   _____

22. Maria says that some multiplication and division fact families have only one multiplication fact and one division fact. Explain why Maria's statement is true. Include examples as part of your explanation.

   _____

   _____

   _____

CCSS: 3.OA.3, 3.OA.8

Lesson 15: Solving Problems with Division

You can use division facts to solve word problems.

Example

Tim has 35 photos from his camping trip. He puts 5 photos on each page of his photo album. How many pages of his album does Tim fill with his camping photos?

Write an equation. Let a ☐ stand for the unknown number of pages.
35 photos ÷ 5 photos on each page = total number of pages filled
$35 \div 5 = $ ☐

Use repeated subtraction to find the unknown number.

$35 - 5 = 30$	$15 - 5 = 10$
$30 - 5 = 25$	$10 - 5 = 5$
$30 - 5 = 20$	$5 - 5 = 0$
$20 - 5 = 15$	

You subtracted 5 from 35 a total of 7 times. $35 \div 5 = 7$

Tim filled 7 pages of his album with photos from his camping trip.

Example

Mr. Barnes has a wooden board that is 8 feet long. He wants to cut the board into 2 equal pieces. How long will each piece of the board be?

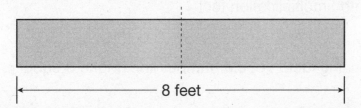

← 8 feet →

Write an equation to help you find the length of each piece of the board.
8 feet ÷ 2 equal pieces = length of each piece
$8 \div 2 = $ ☐

Think of a related multiplication fact.
$2 \times ? = 8$
$2 \times 4 = 8$
So, $8 \div 2 = 4$.

Each piece of the board will be 4 feet long.

Example

Janna has 18 plants. She wants to put 6 plants in each row in her garden. How many rows will there be?

Write an equation to find the number of rows that Janna will plant.
18 plants ÷ 6 plants in each row = total number of rows
18 ÷ 6 = ☐

Use skip-counting to find the unknown number.
Skip count by 6s, the number of plants in each row.
    6, 12, 18
There are 3 groups of 6 in 18.
18 ÷ 6 = 3

There will be 3 rows of plants in Janna's garden.

Example

Kim compares two fruit baskets at the farmer's market. The large fruit basket costs $24. The large fruit basket costs 3 times as much as the small fruit basket. What is the cost of the small fruit basket?

Write an equation to find how much the small fruit basket costs.
$24 ÷ 3 = the cost of the small fruit basket
24 ÷ 3 = ☐

Think of a related multiplication fact.
3 × ☐ = 24

To find the missing factor, use a number line. Make 3 equal jumps.

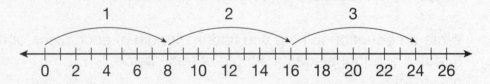

The number of tick marks in each jump is the missing factor.
24 ÷ 3 = 8

The small fruit basket costs $8.

CCSS: 3.OA.3, 3.OA.8

Example

Erik has 48 inches of packaging tape. That is 6 times as much as he needs to wrap a package. How long is the piece of tape that Erik needs for his package?

Write an equation to find how long the piece of tape for Erik's package is.
48 inches ÷ 6 = the length of the piece of tape that Erik needs
48 ÷ 6 = ☐

Think of a related multiplication fact.
   $6 \times ? = 48$
   $6 \times \mathbf{8} = 48$
So, 48 ÷ 6 = 8.

Erik needs 8 inches of tape for his package.

Example

Rosa has two boxes to take on a trip. The larger box weighs 20 pounds. That is 5 times as much as the weight of the smaller box. What is the combined weight of the two boxes?

First, divide to find the weight of the smaller box.
Write an equation. Let *s* equal the weight of the smaller box.
$20 \div 5 = s$

Divide: 20 ÷ 5 = 4, so $s = 4$.

The smaller box weighs 4 pounds.

Then, add to find the combined weight of the two boxes.
Write an equation. Let *p* equal the total pounds that the two boxes weigh.
$20 + 4 = p$

Add: 20 + 4 = 24, so $p = 24$.

The combined weight of the two boxes is 24 pounds.

Practice

1. Teo has 24 model cars. He puts an equal number of the cars in 4 cases. How many cars are in each case?

   total number of cars ÷ number of cases = number of cars in each case

   _____ ÷ _____ = _____

   There are _____ cars in each case.

2. Pedro paid $25 for 5 tubes of paint. Each tube of paint cost the same amount. How much did each tube of paint cost?

   total amount paid ÷ number of tubes = cost of each tube

   _____ ÷ _____ = _____

   Each tube of paint cost _____.

Directions: For questions 3 through 5, write a division equation. Then solve the equation.

3. Amanda has a recipe for crackers. The recipe makes 36 crackers. The recipe makes 4 times as many crackers as she wants. How many crackers does Amanda want to make?

   _____

4. There are 32 desks in Ms. Webb's classroom. There are 8 equal rows of desks. How many desks are in each row?

   _____

5. Zack walked 21 miles in 7 days. He walked the same number of miles each day. How many miles did Zack walk each day?

   _____

6. Marcus has 56 inches of string. He cuts the string into 8 equal pieces. How long is each piece?

    _____

7. Rebecca has 30 carrot sticks to place on 6 small plates. She puts the same number of carrots on each plate. How many carrot sticks does she put on each plate?

    _____

8. Ellis paid $54 for 9 tickets to a talent show. Each ticket cost the same amount. What was the cost of each ticket?

    _____

9. Mr. Cook bought a small sheet of 20 stamps. The sheet of stamps had 5 rows, with the same number of stamps in each row. How many stamps were in each row?

    _____

10. Krista has a plastic pipe that is 15 feet long. She cuts the pipe into 3 equal pieces. How long is each piece?

    _____

11. Tyler spent $8 for 2 sandwiches and $2 on a drink. How much does 1 sandwich and a drink cost?

    _____

12. Melissa talked on the phone for 50 minutes. She talked 10 times as long as her brother. How many more minutes did Melissa talk on the phone than her brother?

    _____

13. Brandon planted a row of lettuce 30 feet long. The row of lettuce was 3 times as long as a row of beets. How many feet long was the row of beets?

   A. 3
   B. 9
   C. 10
   D. 90

14. Serina put 2 cherries on top of each fruit cup. She used a total of 16 cherries. How many fruit cups were there?

   A. 6
   B. 8
   C. 14
   D. 32

15. There are 18 people going on a canoe trip. Each canoe holds 2 people. How many canoes will they need?

   A. 6
   B. 7
   C. 8
   D. 9

16. Masud rode his bike 42 miles. He rode 6 times as far as Ali. How many miles did Ali ride?

   A. 7
   B. 8
   C. 9
   D. 42

17. Mr. Ortiz bought 2 pumpkins. The larger pumpkin weighed 6 times as much as the smaller pumpkin. The larger pumpkin weighed 24 pounds. How much did the 2 pumpkins weigh in all?

   _____

Explain how you found your answer.

_____

_____

_____

Unit 2 Practice Test

1. What multiplication sentence does the array of fish show?

_____ × _____ = _____

2. Desmond sorted his crayons into groups of 5. He has 5 groups. Write a multiplication sentence to show how many crayons Desmond has.

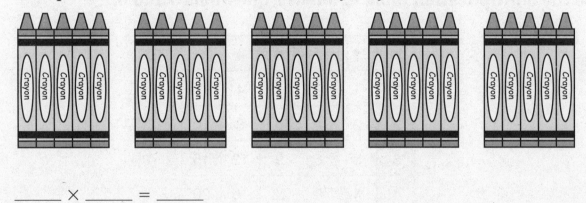

_____ × _____ = _____

3. Sammy has 15 marbles. He wants to divide them into 3 equal groups. How many marbles will there be in each group? Use equal sharing by writing the group number (1, 2, or 3) under each of the following marbles. Then count the number of marbles in each group.

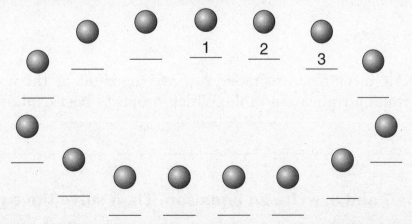

Write a division sentence: _____ ÷ _____ = _____

There will be _____ marbles in each group.

4. Franco says that 3×6 and 6×3 have the same product.

   Is he correct? _____

   Draw a picture to support your answer.

Use the multiplication table to answer questions 5 and 6.

×	1	2	3	4	5
1	1	2	3	4	5
2	2	4	6	8	10
3	3	6	9	12	15
4	4	8	12	16	20
5	5	10	15	20	25

5. The numbers on either side of the diagonal line are the same. Which property best explains why?

   _____

6. The numbers in the first unshaded row are the same as the numbers across the top of the multiplication table. Which property best explains why?

   _____

For questions 7 and 8, write an equation. Then solve the equation.

7. Sherri has 7 packages of stickers. Each package has 6 stickers. How many stickers does Sherri have in all?

   _____

8. Mr. Johnson spent $10 on soup. He bought 5 cans of soup. Each can of soup cost the same amount. What was the cost of each can of soup?

_____

For questions 9 through 20, find the product.

9. $10 \times 4 =$ _____

10. $3 \times 7 =$ _____

11. $4 \times 6 =$ _____

12. $4 \times 2 \times 5 =$ _____

13. $3 \times 10 =$ _____

14. $4 \times 7 =$ _____

15. $8 \times 30 =$ _____

16. $30 \times 3 =$ _____

17. $5 \times 90 =$ _____

18. $4 \times 70 =$ _____

19. $60 \times 7 =$ _____

20. $5 \times 20 =$ _____

For questions 21 through 26, find the quotient.

21. $40 \div 8 =$ _____

22. $28 \div 4 =$ _____

23. $16 \div 2 =$ _____

24. $54 \div 9 =$ _____

25. $6 \div 6 =$ _____

26. $25 \div 5 =$ _____

For questions 27 through 32, find the missing number.

27. $9 \times \square = 18$ _____

28. $\square \div 8 = 8$ _____

29. $4 \times \square = 36$ _____

30. $\square \times 4 = 28$ _____

31. $\square \div 3 = 10$ _____

32. $40 \div \square = 8$ _____

Choose the correct answer.

33. Which multiplication equation does the following picture show?

   A. $5 \times 3 = 15$

   B. $6 \times 2 = 12$

   C. $6 \times 3 = 18$

   D. $3 \times 5 = 15$

34. Rosie and Liz each have the same number of squares in their quilts. Rosie's quilt has 9 rows with 7 squares in each row. Liz's quilt has 7 rows. How many squares are in each row of Liz's quilt?

   A. 7

   B. 9

   C. 18

   D. 63

35. Which group of numbers makes up a multiplication and division fact family?

   A. 4, 6, 24

   B. 4, 7, 27

   C. 6, 9, 63

   D. 8, 9, 64

36. Which multiplication equation does the following model show?

   A. $2 \times 12 = 24$

   B. $3 \times 7 = 21$

   C. $3 \times 8 = 24$

   D. $4 \times 6 = 24$

37. There are 8 students who help Mr. Farley record the weather for his science class. He wants to give each student 4 colored pencils. How many colored pencils does Mr. Farley need?

   A. 2

   B. 12

   C. 24

   D. 32

38. What number can replace the $\square$ to make the equation true?

$$6 \times \square = 6$$

   A. 0

   B. 1

   C. 6

   D. 12

Solve each problem.

39. Habib bought 2 pounds of apples. Justin bought 3 times as many pounds of apples. How many pounds of apples did Justin buy?   _____

40. Carrie is making omelets. She uses 2 eggs to make each omelet. If Carrie has 8 eggs, how many omelets can she make?   _____

41. Jennice has 36 hair clips. She divides them evenly among 4 boxes. How many hair clips does Jennice put in each box?   _____

42. Megan has 4 bags. She puts 6 cookies in each bag. Kate has 2 times as many bags as Megan. Kate also puts 6 cookies in each bag. How many cookies does Kate have?   _____

43. In his room, Rafi has 7 book shelves. Each shelf has 8 books on it. On his desk, he has 5 more books. How many books does Rafi have in all?   _____

44. Nick has 3 aquariums. Each aquarium has 9 fish. How many fish does Nick have in all?   _____

45. Daniella is making bracelets for a craft fair. She plans to make 9 bracelets. She uses 7 beads for each bracelet.

Part A
Write the multiplication equation you could use to find the number of beads that Daniella needs.

_____

Part B
Show how you could use the distributive property to solve the multiplication equation you wrote in Part A.

_____

Part C
Daniella's friend Josie is also making bracelets for the craft fair. Josie uses 6 beads for each bracelet. If Josie makes 10 bracelets for the craft fair, who uses more beads: Daniella or Josie?

_____

How many more beads does the girl you named use for each bracelet?

_____

Explain how you found your answer.

_____

_____

_____

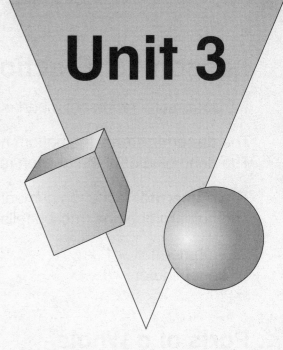

Unit 3

Number and Operations– Fractions

Cooking often involves fractions. To bake a batch of granola, you might use fractions to measure the amounts of ingredients. For example, you might use $\frac{1}{4}$ cup of oil, $\frac{3}{4}$ cup of oats, and $\frac{1}{2}$ cup of nuts. When comparing recipes, you might want to know which recipe uses more flour: Is $\frac{3}{4}$ cup more or less than $\frac{2}{3}$ cup? People in many careers use fractions. For example, carpenters and plumbers use fractions to describe sizes of nails, screws, and pipes, and the thickness of boards.

In this unit, you will work with numbers written as fractions. You will solve problems by comparing fractions.

In This Unit

Fractions

Whole Numbers as Fractions

Equivalent Fractions

Comparing Fractions

CCSS: 3.NF.1, 3.NF.2.a, 3.NF.2.b

Lesson 16: Fractions

A **fraction** can represent a part of a whole or a part of a group of objects.

The **denominator** is the bottom number of a fraction.
The denominator of the fraction tells how many parts the whole is divided into.

The **numerator** is the top number of a fraction.
The numerator of the fraction tells how many parts of the whole you have.

numerator → $\underline{1}$
denominator → 4

Parts of a Whole

To be a fraction, the whole must be divided into **equal** parts. This means that each part must be the same size.

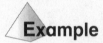 **Example**

What fraction of the window is broken?

The window is divided into 4 equal parts. One of the parts is broken.
So, $\frac{1}{4}$ of the window is broken. In word form, $\frac{1}{4}$ is "one fourth."

TIP: $\dfrac{\text{The n}\textbf{U}\text{merator is } \textbf{U}\text{pstairs}}{\text{The } \textbf{D}\text{enominator is } \textbf{D}\text{ownstairs.}}$

CCSS: 3.NF.1, 3.NF.2.a, 3.NF.2.b

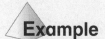**Example**

Which rectangle shows $\frac{1}{3}$ shaded?

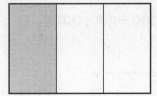 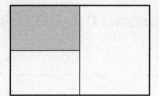

Both of the rectangles are divided into 3 parts and have 1 part shaded. However, the rectangle on the left has 3 equal parts, but the rectangle on the right has 3 parts that are not equal.

The rectangle on the left shows $\frac{1}{3}$ shaded.

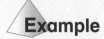**Example**

What fraction of the rectangle is shaded?

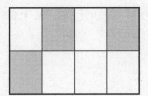

The rectangle is divided into 8 equal parts.
Each part is $\frac{1}{8}$ of the rectangle.

Three of the $\frac{1}{8}$s are shaded. Altogether, $\frac{3}{8}$ are shaded.

So, $\frac{3}{8}$ of the rectangle is shaded.

CCSS: 3.NF.1, 3.NF.2.a, 3.NF.2.b

Fractions on the Number Line

Another way to show fractions is with a number line. The tick marks on a number line are equally spaced, so the number line is divided into equal parts.

This number line shows 0 to 1. It is divided into three equal parts.

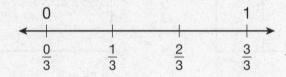

What fraction belongs on the line under the following number line?

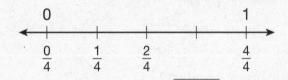

This number line from 0 to 1 is divided into 4 equal parts. Each part is $\frac{1}{4}$.

The fraction $\frac{3}{4}$ belongs on the number line.

What fraction belongs on the line under this number line?

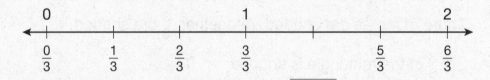

This number line shows 0 to 2. From 0 to 1, it is divided into three equal parts. From 1 to 2, it is also divided into three equal parts.

Count up the thirds: $\frac{1}{3}, \frac{2}{3}, \frac{3}{3}, \frac{4}{3}, \frac{5}{3}, \frac{6}{3}$

The missing fraction is between $\frac{3}{3}$ and $\frac{5}{3}$.

The fraction $\frac{4}{3}$ belongs on the number line.

CCSS: 3.NF.1, 3.NF.2.a, 3.NF.2.b

Practice

1. Circle the figure that shows $\frac{1}{4}$ shaded.

2. Circle the figure that shows $\frac{1}{2}$ shaded.

3. Circle the figure that shows $\frac{1}{6}$ shaded.

4. Circle the figure that shows $\frac{4}{6}$ shaded.

 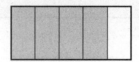 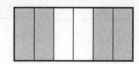

5. Circle the figure that shows $\frac{3}{8}$ shaded.

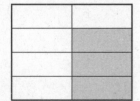  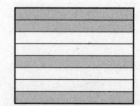

CCSS: 3.NF.1, 3.NF.2.a, 3.NF.2.b

Directions: For questions 6 through 9, write the fraction that represents the shaded part of each figure.

6.  _____

8. 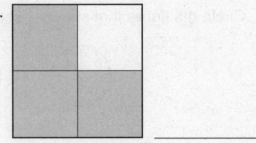 _____

7. 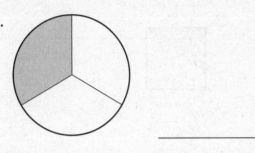 _____

9. 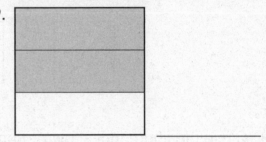 _____

Directions: For questions 10 through 12, write the fraction or fractions that are missing under the number line.

10.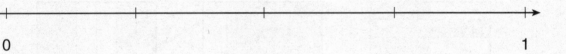

11.

12.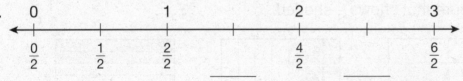

CCSS: 3.NF.1, 3.NF.2.a, 3.NF.2.b

13. Color $\frac{1}{6}$ of the following rectangle.

14. What fraction represents the shaded part of the picture below?

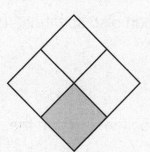

    A. $\frac{1}{4}$

    B. $\frac{1}{2}$

    C. $\frac{2}{3}$

    D. $\frac{3}{4}$

15. Color $\frac{4}{8}$ of the following circle.

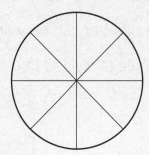

16. What fraction represents the shaded part of the picture below?

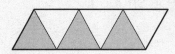

    A. $\frac{1}{3}$

    B. $\frac{2}{5}$

    C. $\frac{3}{6}$

    D. $\frac{3}{7}$

17. Tina says that her drawing shows the fraction $\frac{2}{3}$.

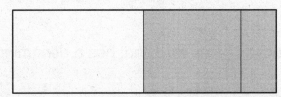

Do you agree? _____

Explain your answer.

_____

_____

111

Lesson 17: Whole Numbers as Fractions

Any whole number can be written as a fraction.

This rectangle is divided into 2 equal sections. Both sections are shaded.

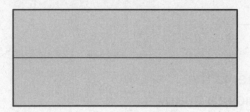

You can say that 1 whole rectangle is shaded. You can also say that $\frac{2}{2}$ of the rectangle is shaded.

This number line shows the number 1 written as a fraction with 2 in the denominator.

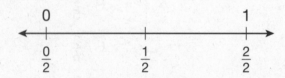

1 and $\frac{2}{2}$ are just different names for the same point on the number line.

A whole number can be written as a fraction with any denominator except 0.

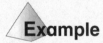 **Example**

What is the fraction name for 1 that has a denominator of 4?

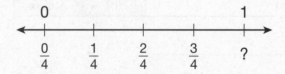

Count up the fourths to find the fraction name for 1: $\frac{0}{4}, \frac{1}{4}, \frac{2}{4}, \frac{3}{4}, \frac{4}{4}$.

The fraction name for 1 that has a denominator of 4 is $\frac{4}{4}$.

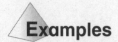
Examples

What is the fraction name for 2 with a denominator of 3?

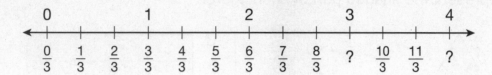

Read it from the number line: 2 and $\frac{6}{3}$ name the same point.
The fraction name for 2 with a denominator of 3 is $\frac{6}{3}$.

What is the fraction name for 3 with a denominator of 3?

Use the number line.

Count up from $\frac{6}{3}$ by $\frac{1}{3}$s to find the fraction name

for 3: $\frac{6}{3}, \frac{7}{3}, \frac{8}{3}, \frac{9}{3}$
The fraction name for 3 with a denominator of 3 is $\frac{9}{3}$.

What is the fraction name for 4 with a denominator of 3?

Use the number line.

Count up from $\frac{9}{3}$ by $\frac{1}{3}$s to find the fraction name

for 4: $\frac{9}{3}, \frac{10}{3}, \frac{11}{3}, \frac{12}{3}$
The fraction name for 4 with a denominator of 3 is $\frac{12}{3}$.

You can write any whole number as a fraction by writing the whole number as the numerator and 1 as the denominator. This is because the fraction bar means to divide the numerator by the denominator.

$\frac{6}{1}$ means $6 \div 1$. $6 \div 1 = 6$

So, the whole number 6 can be written as the fraction $\frac{6}{1}$.

Example

What is the fraction name for 8 that has a denominator of 1?

Think of the division facts: $8 = \boxed{} \div 1$

$$8 = \mathbf{8} \div 1$$

The fraction name for 8 that has a denominator of 1 is $\frac{8}{1}$.

Practice

Directions: For questions 1 through 4, write the fraction and the whole number that represent the shaded part of each figure.

1.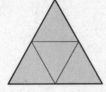

_____

3.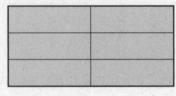

_____

2.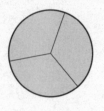

_____

4.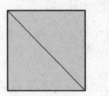

_____

5. What is the fraction name for 1 that has a denominator of 6?

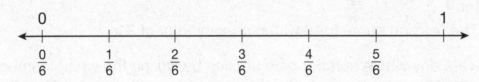

_____

6. What is the fraction name for 3 that has a denominator of 2?

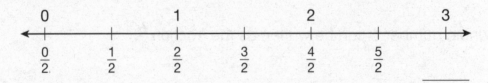

_____

7. What is the fraction name for 1 that has a denominator of 8?

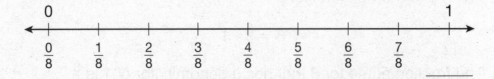

_____

Directions: For questions 8 through 13, write the fraction name with a denominator of 1 for each whole number.

8. 2 = _____

9. 5 = _____

10. 4 = _____

11. 3 = _____

12. 12 = _____

13. 10 = _____

Directions: Use the rectangle to answer questions 14 and 15.

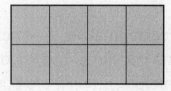

14. Which whole number names the shaded part of the rectangle?

    A. 1

    B. 2

    C. 4

    D. 8

15. Which fraction names the shaded part of the rectangle?

    A. $\frac{1}{8}$

    B. $\frac{2}{8}$

    C. $\frac{4}{8}$

    D. $\frac{8}{8}$

16. Amy wants to show $\frac{5}{6}$ on this number line. What will Amy's number line look like? Complete the number line. Label it to show $\frac{5}{6}$.

Explain how you found the point for $\frac{5}{6}$ on the number line.

_____

_____

CCSS: 3.NF.3.a, 3.NF.3.b

Lesson 18: Equivalent Fractions

Two different fractions can represent the same part of a whole. These are called **equivalent fractions**.

These two circles are the same size. Look at the shaded part of each circle.

$$\frac{1}{2} = \frac{2}{4}$$

The shaded part of the first circle represents the fraction $\frac{1}{2}$. The shaded part of the second circle represents the fraction $\frac{2}{4}$. The shaded parts of the two circles are exactly the same size. Therefore, $\frac{1}{2}$ and $\frac{2}{4}$ are equivalent fractions.

Example

Are $\frac{1}{3}$ and $\frac{2}{6}$ equivalent fractions?

Use fraction bars to find out if the fractions are equivalent.

1					
$\frac{1}{3}$		$\frac{1}{3}$		$\frac{1}{3}$	
$\frac{1}{6}$	$\frac{1}{6}$	$\frac{1}{6}$	$\frac{1}{6}$	$\frac{1}{6}$	$\frac{1}{6}$

Two sections labeled $\frac{1}{6}$ are shaded to show $\frac{2}{6}$.

The bar for $\frac{1}{3}$ is the same length as the bars showing $\frac{2}{6}$.

So, $\frac{1}{3}$ and $\frac{2}{6}$ are equivalent fractions. We say that $\frac{1}{3} = \frac{2}{6}$.

CCSS: 3.NF.3.a, 3.NF.3.b

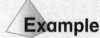 **Example**

Are $\frac{1}{2}$ and $\frac{4}{8}$ equivalent fractions?

Use number lines to find out if the fractions are equivalent.

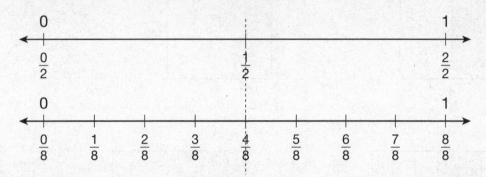

The point for $\frac{1}{2}$ lines up exactly with the point for $\frac{4}{8}$. If one of these number lines were placed on top of the other one, the point for $\frac{4}{8}$ would be the same point as the point for $\frac{1}{2}$.

So, $\frac{1}{2}$ and $\frac{4}{8}$ are equivalent fractions. That is, $\frac{1}{2} = \frac{4}{8}$.

Example

What fraction is equivalent to $\frac{3}{4}$?

Find $\frac{3}{4}$ on the top number line. Read the fraction that is on the matching number line.

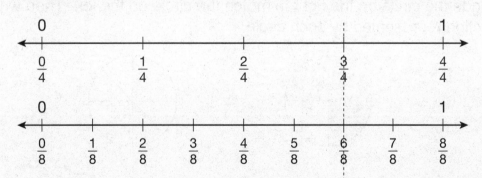

The point for $\frac{3}{4}$ lines up exactly with the point for $\frac{6}{8}$.
So, $\frac{6}{8}$ is equivalent to $\frac{3}{4}$. $\frac{3}{4} = \frac{6}{8}$

117

Practice

Directions: For questions 1 through 3, write an equivalent fraction that represents the shaded part of each figure.

1. 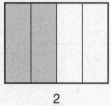 =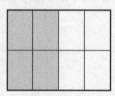

$\frac{2}{4}$ = _____

2. 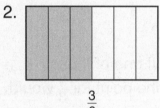 =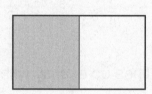

$\frac{3}{6}$ = _____

3 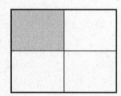 =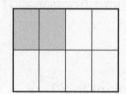

$\frac{1}{4}$ = _____

4. Shade the circle on the right to match the circle on the left. Then write the fraction represented by each circle.

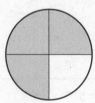 =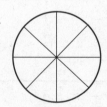

_____ = _____

CCSS: 3.NF.3.a, 3.NF.3.b

Directions: Use the following number lines to answer questions 5 through 9.

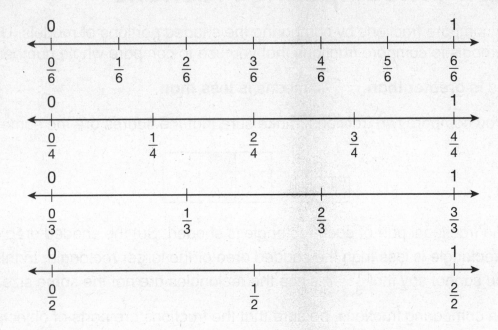

5. What fractions are equivalent to $\frac{1}{2}$? _____ and _____

6. What fraction is equivalent to $\frac{2}{3}$? _____

7. Which fraction is equivalent to $\frac{1}{3}$?

   A. $\frac{2}{6}$

   B. $\frac{3}{1}$

   C. $\frac{3}{3}$

   D. $\frac{6}{2}$

8. Which fraction is equivalent to $\frac{2}{4}$?

   A. $\frac{2}{6}$

   B. $\frac{1}{4}$

   C. $\frac{1}{3}$

   D. $\frac{1}{2}$

9. Nadia says that $\frac{3}{6}$ is equivalent to $\frac{2}{4}$. Do you agree? _____

   Use the number lines to explain your answer.

   _____

   _____

CCSS: 3.NF.3.d

Lesson 19: Comparing Fractions

You can compare fractions by comparing the shaded portions of models. Use the same symbols to compare fractions that you use to compare whole numbers.

$>$ means **is greater than**. $<$ means **is less than**.

Before you compare two fractions, make sure that the figures are the same size.

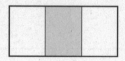

The same fractional part of each rectangle is shaded. But the shaded area of the smaller rectangle is less than the shaded area of the larger rectangle. In this case, you cannot say that $\frac{1}{3} = \frac{1}{3}$ since the rectangles are not the same size.

So, when comparing fractions, be sure that the fractions are parts of objects that are the same size. When fractions have the **same denominators**, compare the numerators. The fraction with the **greater numerator** is the **greater fraction**.

Example

Compare $\frac{2}{4}$ and $\frac{3}{4}$.

$$\frac{2}{4} \qquad\qquad \frac{3}{4}$$

Compare the shaded parts of the two circles. The shaded part of the first circle is smaller than the shaded part of the second circle. So, $\frac{2}{4}$ is less than $\frac{3}{4}$.

You can also use reasoning to compare fractions with the same denominator. The denominators are the same, so compare the numerators. 2 is less than 3, so $\frac{2}{4}$ is less than $\frac{3}{4}$.

$$\frac{2}{4} < \frac{3}{4}$$

CCSS: 3.NF.3.d

When fractions have the **same numerators**, compare their denominators. The fraction with the **lesser denominator** is the **greater fraction**.

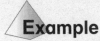 **Example**

Which fraction is greater: $\frac{2}{4}$ or $\frac{2}{8}$?

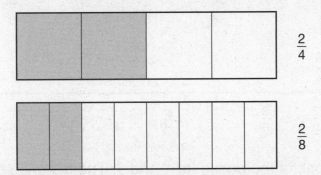

Compare the shaded parts of the fraction strips. The shaded part of the fraction strip for $\frac{2}{4}$ is greater than the shaded part of the fraction strip for $\frac{2}{8}$. So, $\frac{2}{4}$ is greater than $\frac{2}{8}$.

You can also use reasoning to compare fractions with the same numerator. The numerators are the same, so compare the denominators. 4 is less than 8, so $\frac{2}{4}$ is greater than $\frac{2}{8}$.

$\frac{2}{4}$ is the greater fraction.

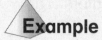**Example**

Compare $\frac{3}{6}$ and $\frac{3}{8}$.

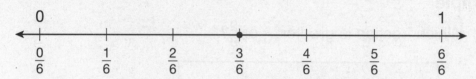

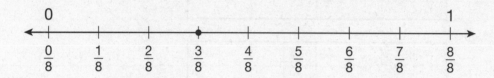

Compare the locations of the points for $\frac{3}{6}$ and $\frac{3}{8}$. The distance of the point for $\frac{3}{6}$ from 0 is greater than the distance of the point for $\frac{3}{8}$ from 0.

So, $\frac{3}{6}$ is greater than $\frac{3}{8}$.

Or, use reasoning. The numerators are the same, so compare the denominators, $6 < 8$. So, $\frac{3}{6}$ is greater than $\frac{3}{8}$.

$\frac{3}{6} > \frac{3}{8}$

Sometimes you can solve problems by comparing fractions.

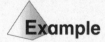**Example**

Riley has $\frac{1}{8}$ pound of peanuts. John has $\frac{3}{8}$ pound of peanuts. Who has more peanuts?

The fractions both have 8 as a denominator. You can compare the numerators to compare the fractions. $1 < 3$, so $\frac{1}{8} < \frac{3}{8}$.

John has more peanuts.

Practice

Directions: For questions 1 and 2, write the fraction represented by the shaded part of each figure. Then use > or < to compare the fractions.

1.

_____   _____

2.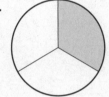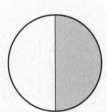

_____   _____   _____

Directions: For questions 3 through 16, write > , < or = to compare the fractions.

3. $\frac{3}{8}$ _____ $\frac{7}{8}$

4. $\frac{5}{8}$ _____ $\frac{5}{6}$

5. $\frac{3}{4}$ _____ $\frac{3}{6}$

6. $\frac{3}{4}$ _____ $\frac{1}{4}$

7. $\frac{2}{3}$ _____ $\frac{2}{6}$

8. $\frac{1}{3}$ _____ $\frac{2}{3}$

9. $\frac{2}{3}$ _____ $\frac{2}{8}$

10. $\frac{2}{8}$ _____ $\frac{2}{6}$

11. $\frac{6}{8}$ _____ $\frac{6}{6}$

12. $\frac{1}{4}$ _____ $\frac{2}{4}$

13. $\frac{5}{6}$ _____ $\frac{3}{6}$

14. $\frac{2}{2}$ _____ $\frac{1}{2}$

15. $\frac{3}{6}$ _____ $\frac{3}{6}$

16. $\frac{3}{3}$ _____ $\frac{3}{8}$

17. Derek ran $\frac{3}{6}$ mile. Pedro ran $\frac{5}{6}$ mile. Who ran the longer distance?

_____

18. Nancy uses $\frac{2}{4}$ pound of grapes and $\frac{2}{8}$ pound of apples to make a fruit salad. Does Nancy use more grapes or more apples to make her fruit salad?

_____

19. Which statement is true?

    A. $\frac{1}{2} < \frac{1}{4}$

    B. $\frac{3}{4} > \frac{2}{4}$

    C. $\frac{2}{3} < \frac{2}{4}$

    D. $\frac{3}{6} > \frac{3}{4}$

20. Which statement is **not** true?

    A. $\frac{2}{3} > \frac{2}{6}$

    B. $\frac{1}{8} < \frac{3}{8}$

    C. $\frac{1}{3} < \frac{1}{4}$

    D. $\frac{4}{6} > \frac{3}{6}$

21. Mrs. Pioli makes two pans of lasagna. The pans are different sizes. She cuts each pan of lasagna into 8 pieces.

 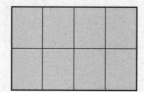

Tony and his brother Billy are each served a piece of lasagna. Tony complains that his piece is smaller than his brother's piece. Billy says that each piece is $\frac{1}{8}$ of a pan, so the pieces must be the same size. Do you agree?

_____

Explain your answer.

_____

_____

_____

Unit 3 Practice Test

For questions 1 through 4, write the fraction that represents the shaded part of each figure.

1.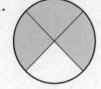

_____

2.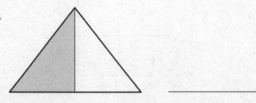

_____

3.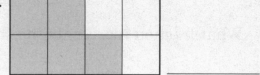

_____

4.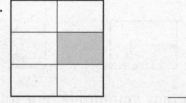

_____

For questions 5 through 7, count the number of equal parts on each number line. Then, write the correct fraction on the blank below the tic mark on the number line.

5.

6.

7.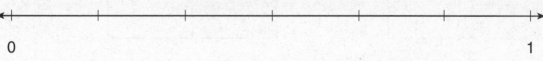

8. Divide the rectangle below into 4 equal parts. Then shade one of the parts.

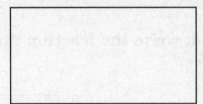

What fraction represents the shaded part of the rectangle? _____

9. Circle the two figures below that represent equivalent fractions.

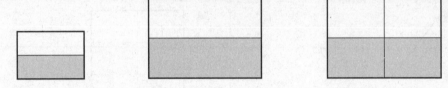

Explain how you found your answer.

_____

_____

For questions 10 and 11, write an equivalent fraction that represents the shaded parts of each figure.

10. 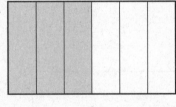 =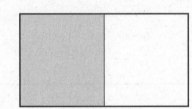

$\frac{3}{6}$ _____

11. 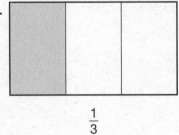 =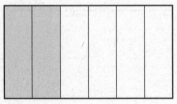

$\frac{1}{3}$ _____

Use the following number lines to answer questions 12 through 17.

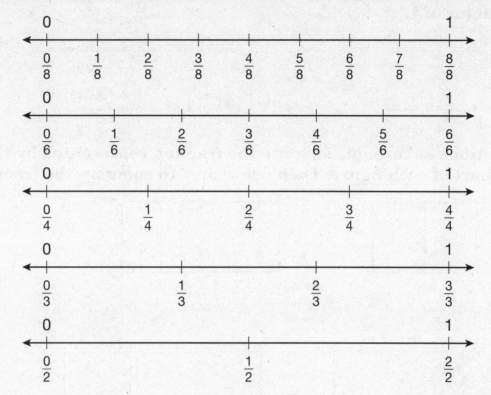

12. What fraction is equivalent to $\frac{2}{3}$? _____

13. What three fractions are equivalent to $\frac{1}{2}$? _____ and _____ and _____

14. What is the fraction name for 1 that has a denominator of 4? _____

15. What is the fraction name for 1 that has a denominator of 2? _____

16. What fraction is equivalent to $\frac{1}{4}$? _____

17. What fraction is equivalent to $\frac{6}{8}$? _____

For questions 18 through 21, write the fraction name with a denominator of 1.

18. 3 = _____

19. 6 = _____

20. 2 = _____

21. 12 = _____

For questions 22 through 24, write the fraction represented by the shaded part of each figure. Then use > or < to compare the fractions.

22.

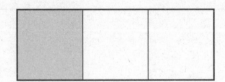

_____ _____ _____

23.

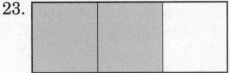

_____ _____ _____

24.

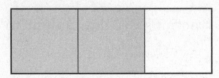

_____ _____ _____

For questions 25 through 30, write >, <, or = to compare the fractions.

25. $\frac{1}{4}$ _____ $\frac{1}{8}$

26. $\frac{4}{6}$ _____ $\frac{4}{6}$

27. $\frac{1}{3}$ _____ $\frac{2}{3}$

28. $\frac{2}{3}$ _____ $\frac{2}{4}$

29. $\frac{7}{8}$ _____ $\frac{5}{8}$

30. $\frac{1}{4}$ _____ $\frac{2}{4}$

31. Which figure shows $\frac{1}{4}$ shaded?

A.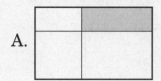

B.

C.

D.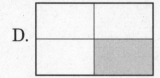

32. Which fraction is least?

A.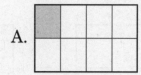

B.

C.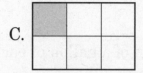

D.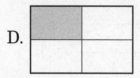

33. Julie wants to show Ed the fraction $\frac{1}{4}$ on the number line. Complete the number line and label it to show $\frac{1}{4}$.

0   1

Explain how you found the point for $\frac{1}{4}$ on the number line.

_____

_____

_____

34. Cindy makes a fruit punch. She has gallon containers of apple juice, lemon juice, and orange juice. She uses part of each gallon as shown below.

apple juice   lemon juice   orange juice

Part A
What fraction of a gallon of each juice does Cindy use?

apple juice _____

lemon juice _____

orange juice _____

Part B
Of which type of juice does she use the greatest amount?

_____

Part C
Cindy decides to add $\frac{1}{4}$ gallon of lemon-lime soda to her punch. Is there more lemon-lime soda or more apple juice in the punch?

_____

Explain how you found your answer.

_____

_____

_____

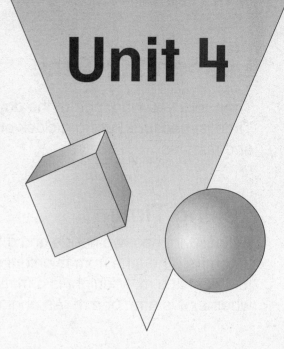

Unit 4

Measurement and Data

Measuring is not just using a ruler to determine the length of a board. You measure time when you look at a clock or a calendar. Every trip to the grocery store involves many different measurements. Which size can of soup holds the most? How many pounds of apples do you need to make two apple pies? Understanding area and perimeter can help you make the best use of a roll of fencing for the garden or decide the best size rug to place in your bedroom.

Graphs are useful tools for presenting information. The data becomes a picture instead of just numbers. Graphs are used daily in magazines and newspapers to present important information so that it is easier to understand.

In this unit, you will measure time using both analog and digital clocks. You will measure length, mass, and capacity. You will find the perimeter of plane figures and the area of rectangles. You will also organize, display, and answer questions about data.

In This Unit

Time

Mass

Capacity

Measuring Length with
  an Inch Ruler

Perimeter

Understanding Area

Area of Rectangles

Rectangles as Models

Irregular Figures

Relating Perimeter and
  Area

Picture Graphs

Bar Graphs

Line Plots

Lesson 20: Time

Time tells you what part of the day it is or how long it takes for an event to occur. Time is measured using a clock or a calendar. A clock measures time in hours and minutes.

Telling Time

A clock that has two hands and a face is an **analog clock.** Each day has 24 hours. The time from midnight until noon is a.m. The time from noon until midnight is p.m. Midnight is a.m. Noon is p.m. An analog clock does not show whether it is a.m. or p.m. An analog clock is shown below.

The short hand of the analog clock points to the hours. The numbers around the clock show the hours.

The long hand points to the minutes. The little marks around the clock show the minutes. There are 5 minutes from one number to the next. There are 60 minutes in one hour, and 30 minutes in a half-hour. This clock shows 8:15.

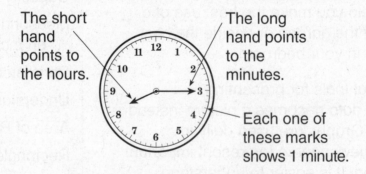

The short hand points to the hours.

The long hand points to the minutes.

Each one of these marks shows 1 minute.

If the time is between midnight and noon, it would be 8:15 a.m. If the time is between noon and midnight, it would be 8:15 p.m.

CCSS: 3.MD.1

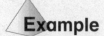

Example

What time is shown on the following clock? It is between noon and midnight.

The short hand shows the hours. It is between the 1 and 2, so the hour is 1.

The long hand shows the minutes. It is between 8 and 9. Because each number represents 5 minutes, multiply 8 by 5 and add the number of marks past the 8 until you get to the long hand. It is 1 mark past the 8.

$8 \times 5 = 40$

$40 + 1 = 41$

Because it is between noon and midnight, the clock shows that the time is 1:41 p.m.

A clock that does not have hands and only shows the current time is a **digital clock**. A digital clock usually has a little light beside the time that shows whether it is a.m. or p.m. A digital clock is shown below.

Hour   Minutes

Since the little light by the a.m. is on, it is 8:15 a.m.

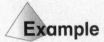

Example

What time is it?

The little light by the p.m. is on, so it is 3:47 p.m.

Elapsed Time

Elapsed time is the amount of time that has passed between two given times.

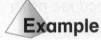

Example

Jamie went outside at 10:35 a.m. She came back inside at 11:15 a.m. How long was Jamie outside?

Start at 10:35. Count the minutes to 11:15.

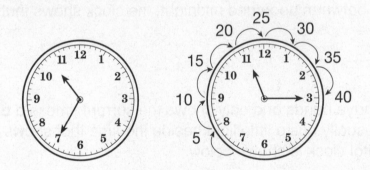

You can also use a number line. Count the minutes by skip counting by 5s.

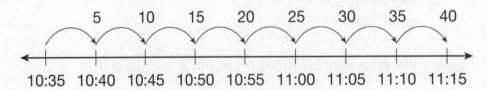

Jamie was outside for 40 minutes.

CCSS: 3.MD.1

Solving Problems with Time

 **Example**

Melanie had a swimming lesson after school. The lesson lasted for 20 minutes. After the lesson, Melanie practiced swimming for 15 more minutes and then left the pool. For how many minutes was Melanie in the pool?

Use a number line to count the minutes that Melanie was in the pool.

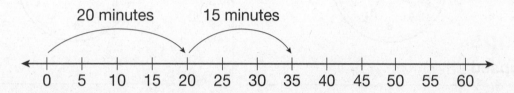

Or, you can add to find the total number of minutes that Melanie was in the pool.

20 + 15 = 35

Melanie was in the pool for 35 minutes.

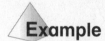 **Example**

Leo spent a total of 50 minutes on homework. He spent 10 minutes on spelling homework. The rest of the time was spent on math. How many minutes did Leo spend on math homework?

Use a number line. Count back from 50 minutes.

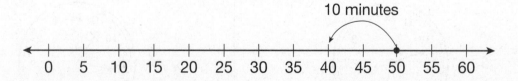

Or, you can subtract the minutes spent on spelling from the total time spent doing homework.

50 − 10 = 40

Leo spent 40 minutes on math homework.

Practice

Directions: For questions 1 through 6, write the time shown on the clock. Be sure to include a.m. or p.m.

1. It is between midnight and noon.

_____

4. It is between noon and midnight.

_____

2. It is between noon and midnight.

_____

5. It is between noon and midnight.

_____

3. It is between midnight and noon.

_____

6. It is between midnight and noon.

_____

CCSS: 3.MD.1

Directions: For questions 7 through 12, write the time shown on the clock. Be sure to include a.m. or p.m.

7.

_____

10.

_____

8.

_____

11.

_____

9.

_____

12.

_____

13. Draw the time 8:25 p.m. on both of the clocks below.

14. It is 6:20 p.m.

What time will it be 15 minutes later? _____

15. Jared's clock shows the time.

What time was it 10 minutes earlier? _____

16. At Frank's school, the bell rings to start lunch at 11:45 a.m. The bell rings to end lunch at 12:17 p.m.

How long is lunch at Frank's school? _____

17. After dinner, Dylan spent 30 minutes reading a book. Then he spent 15 minutes doing a word search puzzle. For how many minutes was Dylan busy with these two activities?

_____

18. Kayla had a piece of lumber. She wanted to build some shelves. She spent 35 minutes deciding how long the shelves should be. She spent 20 minutes measuring and cutting the wood. How many more minutes did she spend planning the shelves than she spent measuring and cutting the wood?

_____

CCSS: 3.MD.1

19. The clock below shows the time when the sun rose yesterday.

What time did the sun rise?

A. 1:30 a.m.

B. 6:01 a.m.

C. 6:05 a.m.

D. 6:05 p.m.

20. Edgar gets home from school at the time shown on the clock below. It takes him 15 minutes to walk home.

At what time does he leave school?

A. 3:05 p.m.

B. 3:20 p.m.

C. 3:35 p.m.

D. 3:50 p.m.

21. Angela went to the dance studio at 3:30 p.m. She spent 40 minutes practicing her tap routine and 10 minutes choosing a costume.

At what time did Angela leave the dance studio? _____

Explain how you found your answer.

_____

_____

_____

Lesson 21: Mass

Mass is a measure of the amount of matter in an object. You can use a balance or a scale to measure mass.

Grams (g) and **kilograms (kg)** are
the metric units used to measure mass.
Grams are used to measure the mass of
light objects.
1 gram
A paper clip has a mass of about 1 gram.

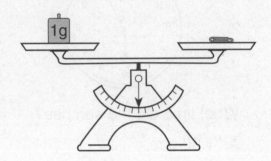

Kilograms are used to measure
the mass of heavier objects.
1 kilogram = **1000 grams**
A tape dispenser has a mass of about
1 kilogram.

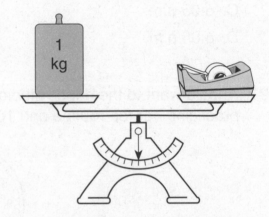

It is important to use an appropriate unit when measuring. Since there are
1000 grams in a kilogram, a gram is a smaller unit than a kilogram.

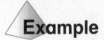 **Example**

What metric unit would be best to measure the mass of a baby?
Why would you choose that unit?

Kilograms would be best to measure the mass of a baby.
Grams are too small. It would take a great many of them to measure
the mass of a baby.

You can solve problems using metric units of mass.

 **Example**

Torrey measured the mass of two dogs, Spot and King.

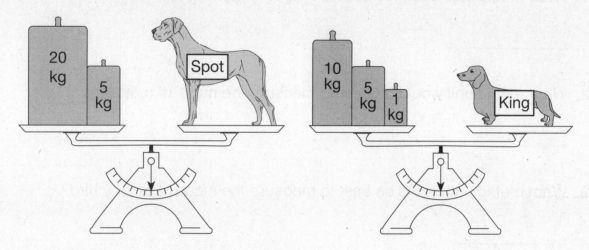

What is the mass of each dog?

Add to find the mass of each dog.

Spot's mass: 20 + 5 = 25
King's mass: 10 + 5 + 1 = 16

Spot has a mass of 25 kilograms. King has a mass of 16 kilograms.

How much greater is Spot's mass than King's mass?

Subtract to find the difference.

25 − 16 = 9

Spot's mass is 9 kilograms greater than King's mass.

Practice

Directions: For questions 1 through 4, write *grams* or *kilograms*.

1. What metric unit would be best to measure the mass of a leaf?

   _____

2. What metric unit would be best to measure the mass of a dictionary?

   _____

3. What metric unit would be best to measure the mass of a baby bird?

   _____

4. What metric unit would be best to measure the mass of a box of apples?

   _____

5. Which real-world object would most likely have a mass of 15 grams? Circle the correct answer.

   folding chair   CD   bowling ball

6. If you were to measure the mass of a TV, which would give you a greater number of units: grams or kilograms?

   _____

   Explain your answer.

   _____

   _____

CCSS: 3.MD.2

Directions: For questions 7 through 9, determine the mass of each object shown on the balance scale.

7.

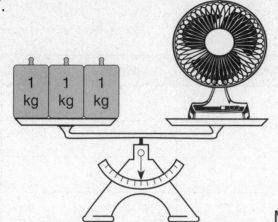

Measurement: _____ kg

8.

Measurement: _____ g

9.

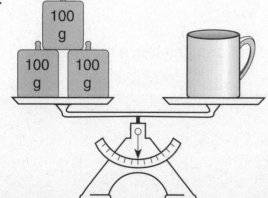

Measurement: _____ g

Directions: For questions 10 through 15, fill in the table with the names of real-world objects. First, estimate the mass of each object. Then measure its mass in metric units. When you write each estimate and measurement, be sure to include the name of the metric units.

	Object	Estimate	Measurement
10.			
11.			
12.			
13.			
14.			
15.			

16. Mrs. Barnes measured the mass of a small box of oranges.

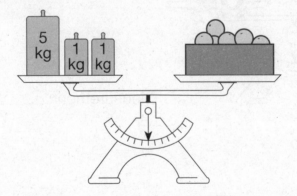

The storekeeper tells her that a large box of oranges has a mass that is 5 kilograms greater.

What is the mass of the large box of oranges? _____ kg

17. Which is the best estimate for the mass of a feather?

    A.   1 gram

    B.   4 kilograms

    C. 15 kilograms

    D. 100 grams

18. Which is the best estimate for the mass of a horse?

    A.   2 kilograms

    B. 145 grams

    C. 450 kilograms

    D. 1000 grams

Directions: Use the picture and other information to answer questions 19 and 20.

Ricardo is going on a trip. He measured the mass of his suitcase.

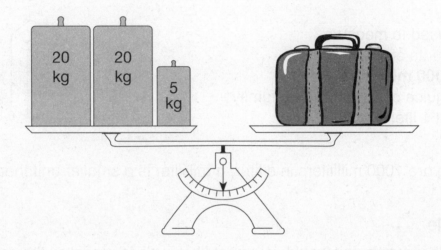

19. What is the mass of Ricardo's suitcase? _____ kg

20. His suitcase cannot have a mass greater than 32 kilograms. What mass does he need to take out of his suitcase?

          _____

Explain how you found your answer.

_____

_____

_____

Lesson 22: Capacity

Capacity is a measure of how much liquid a container can hold.

Milliliters (mL) and **liters (L)** are two metric units used to measure capacity.

Milliliters are used to measure
small amounts of liquid
1 milliliter
A medicine dropper holds about 1 milliliter.

Liters are used to measure
larger amounts of liquid.
1 liter = 1000 milliliters
An orange juice container for the family
holds about 1 liter.

Since there are 1000 milliliters in a liter, a milliliter is a smaller unit than a liter.

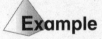 **Example**

What metric unit would you most likely use to measure the capacity of a bottle cap?

A bottle cap holds a small amount of liquid. You would most likely use milliliters to measure its capacity.

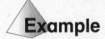 **Example**

What metric unit would you most likely use to measure the capacity of a bucket?

Liters would be better to measure the capacity of a bucket. It would take thousands of drops from a medicine dropper to fill a bucket, so milliliters are too small a unit to use.

CCSS: 3.MD.2

You can solve problems using metric units of capacity.

 **Example**

Jenna has a recipe for punch. The recipe calls for orange juice, pineapple juice, and water. Jenna has measured out the ingredients.

Orange juice   Pineapple juice   Water

How much of each ingredient does Jenna have?

Read the measurements on each container.

Jenna has 2 liters of orange juice, 1 liter of pineapple juice, and 2 liters of water.

How much punch does the recipe make?

Add to find the total amount of punch.

$2 + 1 + 2 = 5$

The recipe makes 5 liters of punch.

 **Example**

Jackson has a 72-milliliter bottle of medicine for his dog. A dose of the medicine is 8 milliliters. How many doses of medicine are in the bottle?

Divide to find the number of doses.

$72 \div 8 = 9$

There are 9 doses of medicine in the bottle.

Practice

Directions: For questions 1 through 3, write how much liquid is shown in each figure.

1.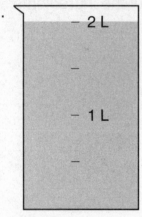

       – 2 L

       –

       – 1 L

       –

Measurement: _____ L

2. – 2000 mL

   –

   – 1000 mL

   –

Measurement: _____ mL

3. – 2 L   – 2 L

   –   –

   – 1 L   – 1 L

   –   –

Measurement: _____ L

CCSS: 3.MD.2

Directions: For questions 4 through 7, write *liter* or *milliliter*.

4. Which metric unit would be best to measure the capacity of a gasoline can?

   _____

5. Which metric unit would be best to measure the capacity of a coffee mug?

   _____

6. Which metric unit would be best to measure the capacity of a fishing pond?

   _____

7. Which metric unit would be best to measure the capacity of a soup bowl?

   _____

8. Which real-world object would most likely have a capacity of 3 liters? Circle the correct answer.

glue bottle   chocolate shake   fish bowl

9. Which is the best estimate for the capacity of a kitchen sink?

   A. 1 liter

   B. 5 liters

   C. 20 liters

   D. 100 liters

10. Which is the best estimate for the capacity of a bathtub?

   A. 1 liter

   B. 5 liters

   C. 10 liters

   D. 60 liters

Directions: For questions 11 through 14, fill in the table with the names of real-world objects. First, estimate the capacity of each object. Then measure its capacity in metric units. When you write each estimate and measurement, be sure to include the name of the metric units.

	Object	**Estimate**	**Measurement**
11.			
12.			
13.			
14.			

15. A restaurant has three full containers of oil on hand for cooking.

Cooking Oil

5 liters

Cooking Oil

5 liters

Cooking Oil

5 liters

How much oil does the restaurant have in the containers? _____

16. Coach Arthur has a 16-liter container of water for the track team. There are 8 members of the track team. If the water is shared equally, how many liters of water can each team member have?

_____

Explain how you found your answer.

_____

_____

CCSS: 3.MD.4

Lesson 23: Measuring Length with an Inch Ruler

An inch ruler is divided into inches and fractions of an inch. The figure below shows $\frac{1}{4}$ and $\frac{1}{2}$ of an inch marked on an inch ruler.

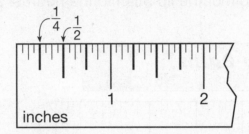

You can measure to the nearest half-inch. You need to look at which $\frac{1}{2}$-inch mark the object is closest to. Sometimes it will be a $\frac{1}{2}$-inch and sometimes it will be a whole number.

Example

How long is this pencil to the nearest $\frac{1}{2}$ inch?

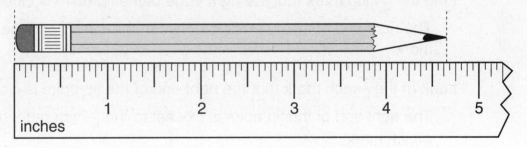

Line up the left end of the ruler with the left end of the pencil.

Count the number of full inches.
 4 full inches

Look at the $\frac{1}{2}$ inch mark that the right end of the pencil is closest to.
 The right end of the pencil is closest to the $\frac{1}{2}$-inch mark after the 4-inch mark.

The length of the pencil to the nearest $\frac{1}{2}$ inch is $4\frac{1}{2}$ inches.

You can measure to the nearest quarter-inch. You need to look at which $\frac{1}{4}$-inch mark the object is closest to. It may be $\frac{1}{4}$-inch, $\frac{1}{2}$-inch, $\frac{3}{4}$-inch, or a whole number of inches.

Example

What is the length of the lip balm to the nearest $\frac{1}{4}$ inch?

Line up the left end of the ruler with the left end of the lip balm.

Count the number of full inches.
   2 full inches

Find the $\frac{1}{4}$-inch mark that the right edge of the lip balm is closest to.
   The right end of the lip balm is between the $\frac{1}{2}$-inch and the $\frac{3}{4}$-inch mark.

Look at the $\frac{1}{4}$-inch mark that the right end of the lip balm is closest to.
   The right end of the lip balm is closest to the $\frac{3}{4}$-inch mark after the 2-inch mark.

The length of the lip balm to the nearest $\frac{1}{4}$ inch is $2\frac{3}{4}$ inches.

Practice

1. In the space below, draw a line that is $3\frac{1}{2}$ inches long.

CCSS: 3.MD.4

Directions: For questions 2 through 5, use an inch ruler to measure each object to the nearest given unit.

2. $\frac{1}{2}$ inch

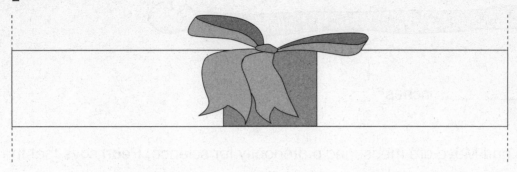

_____ inches

3. $\frac{1}{4}$ inch

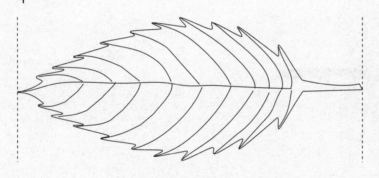

_____ inches

4. $\frac{1}{4}$ inch

_____ inches

5. $\frac{1}{2}$ inch

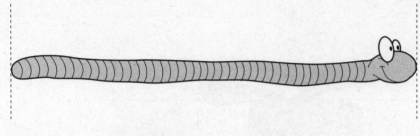

_____ inches

6. Pearl and Maya are measuring a dragonfly for science. Pearl says that the dragonfly measures $1\frac{1}{2}$ inches. Maya disagrees and says that it measures $1\frac{1}{4}$ inches.

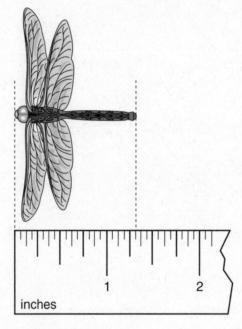

Which student do you agree with?

_____

Explain your answer.

_____

_____

_____

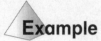Lesson 24: Perimeter

The distance around the outside of a plane figure is called the **perimeter**. To find the perimeter of a polygon, add the lengths of all the sides of the polygon.

**Example**

What is the perimeter of this rectangle?

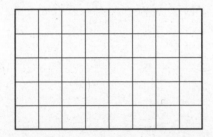

Count the units on each side of the rectangle.
There are 8 units across the top.
There are 5 units along the right side.
There are 8 units across the bottom.
There are 5 units along the left side.

Now, add the lengths of all the sides.

8 + 5 + 8 + 5 = 26

The perimeter of the rectangle is 26 units.

Example

What is the perimeter of the following triangle?

11 inches
7 inches   6 inches

Add the lengths of all the sides.

11 + 7 + 6 = 24

The perimeter of the triangle is 24 inches.

Example

Danny's deck is shown below. What is the perimeter of Danny's deck?

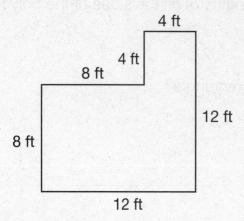

Add the lengths of all the sides.

4 + 12 + 12 + 8 + 8 + 4 = 48

The perimeter of Danny's deck is 48 feet.

Example

What is the perimeter of the following square?

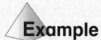 3 yards

The drawing only shows the length of one side. Since the polygon is a square, all the sides are the same length. Add 3 four times.

3 + 3 + 3 + 3 = 12

Since you add the same number 4 times to find perimeter of a square, you can multiply by 4 to find the perimeter.

4 × 3 = 12

The perimeter of the square is 12 yards.

CCSS: 3.MD.8

When all the sides of a figure are the same length, you can find the perimeter by multiplying the number of sides times the length of each side.

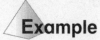 **Example**

Bettina made a hexagon-shaped sign out of cardboard. Each side of the hexagon is 8 inches long. What is the perimeter of Bettina's sign?

8 in.

Six sides are the same length, so multiply the length by 6.
$6 \times 8 = 48$

The perimeter of Bettina's sign is 48 inches.

When you know the perimeter of a figure, you can find a missing measure.

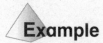 **Example**

The perimeter of a play area is 27 meters. What is the missing side length?

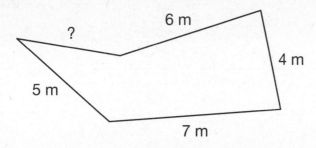

6 m

?

4 m

5 m

7 m

Find the sum of the known side lengths.
$6 + 4 + 7 + 5 = 22$

Subtract the sum of the known side lengths from the perimeter to find the unknown length.
$27 - 22 = 5$

The missing side length is 5 meters.

Practice

Directions: For questions 1 through 4, find the perimeter of each figure.

1.

_____ + _____ + _____ + _____ = _____ units

2.

9 in.   9 in.

9 in.

_____ + _____ + _____ = _____ inches

3.

5 m

2 m

3 m

6 m

_____ + _____ + _____ + _____ = _____ meters

4.

4 cm

4 cm   4 cm

4 cm

_____ × _____ = _____ centimeters

CCSS: 3.MD.8

5. A piece of construction paper is shown below.

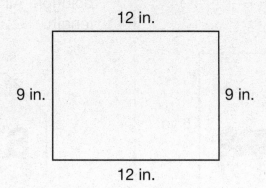

12 in.

9 in.   9 in.

12 in.

What is the perimeter of the piece of paper? _____

6. Darby's playpen is shown below.

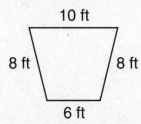

10 ft

8 ft   8 ft

6 ft

What is the perimeter of Darby's playpen? _____

7. A square flower garden is 5 feet on each side. What is the perimeter of the flower garden?

_____

8. A swimming pool is in the shape of a rectangle. It is 3 meters wide and 5 meters long. What is the perimeter of the pool?

_____

9. Josh wants to put a fence around his backyard. His backyard is 7 yards long and 6 yards wide. How many yards of fencing does Josh need?

_____

10. A farmer built this pen for his cows.

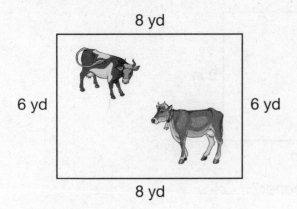

8 yd

6 yd   6 yd

8 yd

What is the perimeter of the pen?

A. 14 yards

B. 24 yards

C. 28 yards

D. 32 yards

11. A stop sign is shaped like an octagon. All 8 sides are the same length.

20 in.

What is the perimeter of this stop sign?

A.   16 inches

B.   40 inches

C.   80 inches

D. 160 inches

12. The perimeter of the figure below is 67 centimeters.

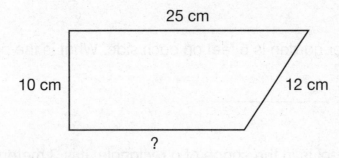

25 cm

10 cm   12 cm

?

What is the missing side length?

_____

Explain how you found your answer.

_____

_____

_____

CCSS: 3.MD.5.a, 3.MD.5.b, 3.MD.6

Lesson 25: Understanding Area

The **area** of a plane figure is the number of **square units** needed to cover the surface of the figure. The square units can be squares with sides of any length. They may be squares with side lengths of 1 inch, 1 foot, 1 centimeter, 1 meter, or any other side length.

To find the area of a plane figure, you can count the number of square units that cover the figure without gaps or overlapping.

 Example

What is the area of this square?

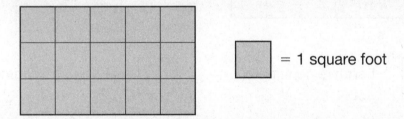

Count the number of square units that make up the square.
   There are 4 small squares.

The area of the square is 4 square units.

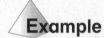 **Example**

What is the area of this rectangle?

Count the number of square feet that make up the rectangle.
   There are 15 small squares.

The area of the rectangle is 15 square feet.

CCSS: 3.MD.5.a, 3.MD.5.b, 3.MD.6

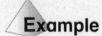

Example

What is the area of this figure?

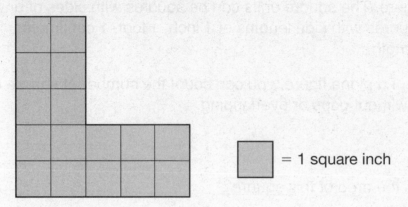

= 1 square inch

Count the number of square inches that make up the figure.
There are 16 small squares.

The area of the figure is 16 square inches.

Example

Zoe drew the first figure. Rose drew the second figure. Which figure has the greater area?

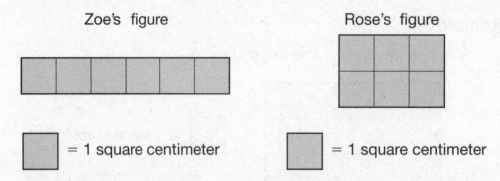

Zoe's figure

Rose's figure

= 1 square centimeter   = 1 square centimeter

Count the number of square centimeters that make up each figure.
Zoe's figure has 6 square centimeters inches.
Rose's figure has 6 square centimeters inches.

The area of each figure is 6 square centimeters inches, so the areas are the same.

162

CCSS: 3.MD.5.a, 3.MD.5.b, 3.MD.6

Practice

Directions: For questions 1 through 4, find the area of each figure in square units.

1.

The area is _____ square units.

2.

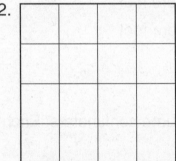

The area is _____ square units.

3.

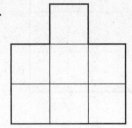

The area is _____ square units.

4.

The area is _____ square units.

Directions: For questions 5 and 6, find the area of each figure.

5.

= 1 square meter

The area is

_____ square _____.

6.

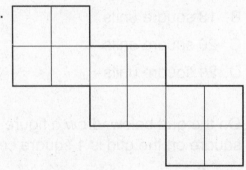

= 1 square inch

The area is

_____ square _____.

163

7. What is the area of the figure?

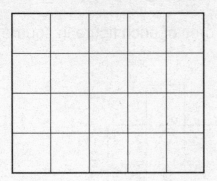

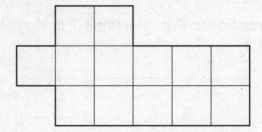

= 1 square unit

A. 16 square units

B. 18 square units

C. 20 square units

D. 24 square units

8. What is the area of the figure?

= 1 square foot

A. 10 square feet

B. 11 square feet

C. 12 square feet

D. 13 square feet

9. On the grid below, draw a figure with an area of 8 square centimeters. Each square on the grid is 1 square centimeter.

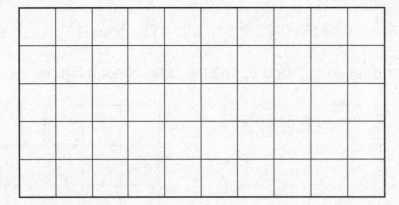

Explain how to find the area of your figure.

_____

_____

_____

CCSS: 3.MD.7.a, 3.MD.7.b

Lesson 26: Area of Rectangles

One way to find the area of a rectangle is by counting the number of square units needed to cover the rectangle. Another way is to think of the square units as a multiplication array.

Remember, you can show multiplication using an array. To multiply 2 × 3, draw an array that is 2 rows by 3 columns.

This array of squares shows that 2 × 3 = 6.
This array is also called an area model.

You can use what you know about area models to find the area of a rectangle.

Example

What is the area of this rectangle?

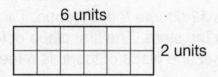

6 units

2 units

Multiply to find the area.

2 × 6 = 12

The area of the rectangle is 12 square units.

You can also count the total number of squares. There are 12 squares, so the area of the rectangle is 12 square units. You get the same answer either way.

CCSS: 3.MD.7.a, 3.MD.7.b

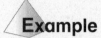 Example

Sheldon is making a square garden bed. The garden will be 5 feet on each side. What will be the area of the garden?

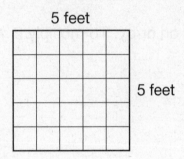

5 feet

5 feet

Multiply to find the area.

$5 \times 5 = 25$

The area of the garden is 25 square feet.

Example

Ms. Cheng needs 32 square feet of fabric. There are two rectangular pieces of fabric in the store. The first piece of fabric is 8 feet long and 4 feet wide. The second piece of fabric is 6 feet long and 5 feet wide.

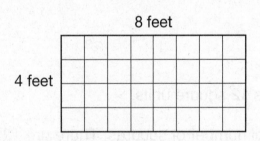

8 feet

4 feet

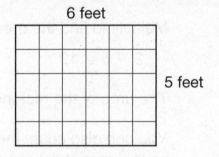

6 feet

5 feet

Which piece of fabric should Ms. Cheng buy?

Find the area of each piece of fabric.

Area of first piece of fabric: $4 \times 8 = 32$
Area of second piece of fabric: $5 \times 6 = 30$

Ms. Cheng needs 32 square feet of fabric, so she should buy the first piece of fabric.

CCSS: 3.MD.7.a, 3.MD.7.b

Practice

Directions: For questions 1 through 6, find the area of each figure.

1.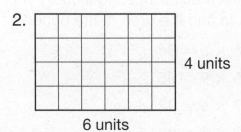

7 units

The area is _____ square units.

2.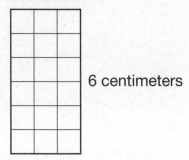

The area is _____ square units.

3. 3 centimeters

The area is _____
square centimeters.

4.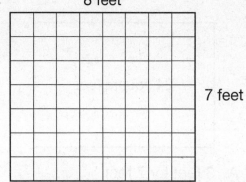

The area is _____ square feet.

5. 4 meters

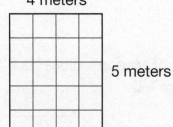

The area is _____
square meters.

6. 7 inches

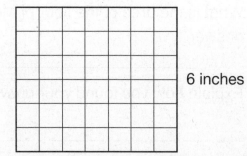

The area is _____ square inches.

7. Mrs. James has a rectangular dining room table. The area of the table top is 24 square feet. Which could be the dimensions of the table?

24 square feet

A. 3 feet by 7 feet

B. 4 feet by 6 feet

C. 4 feet by 8 feet

D. 5 feet by 6 feet

8. Kevin has a rectangular piece of grid paper. Each square is 1 square inch.

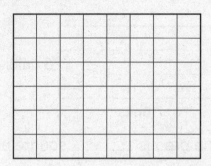

Which multiplication fact could you use to find the area of the paper?

A. $6 \times 6 = 36$

B. $5 \times 9 = 45$

C. $7 \times 6 = 42$

D. $8 \times 6 = 48$

9. The floor of Megan's closet is 8 feet by 5 feet.

8 feet

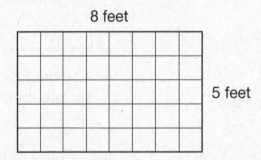

5 feet

What is the area of the floor of Megan's closet? Be sure to label your answer.

_____

Explain how you found your answer.

_____

_____

CCSS: 3.MD.7.c

Lesson 27: Rectangles as Models

You can use rectangles as models to help you understand the distributive property.

The **distributive property** says that multiplying the sum of two numbers by a third number is the same as multiplying each addend by the third number and adding the products.

Here is an example of the distributive property.

$$2 \times (5 + 3) = (2 \times 5) + (2 \times 3)$$
$$2 \times 8 = 10 + 6$$
$$16 = 16$$

This shows that you can break apart a factor into a sum, so you have smaller numbers to multiply: $2 \times 8 = 2 \times (5 + 3)$. The factor 8 is broken into the sum $5 + 3$.

Example

Use a rectangle area model to show how to use the distributive property to multiply 2×8. Explain what the model shows.

Draw a rectangle area model to show 2×8. Then, separate the model into two parts.

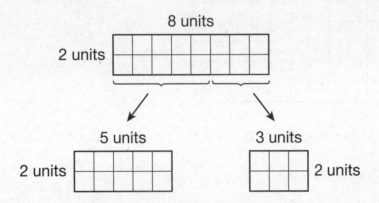

The first model shows that $2 \times 8 = 16$.

The model can be separated into two parts that show $2 \times 5 = 10$ and $2 \times 3 = 6$. The sum of the areas of the two parts ($10 + 6 = 16$) is the same as the product of the first area model, 16.

You can also use the distributive property to combine two multiplication expressions into one expression.

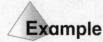

Example

Use rectangle area models and the distributive property to show this multiplication.

$$(6 \times 3) + (6 \times 7)$$

Show a rectangle area model for each multiplication.

$6 \times 3 = 18$ $6 \times 7 = 42$

Combine the two models to make one rectangle area model.

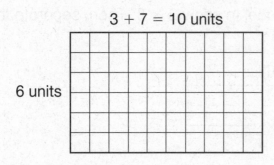

The models show that $(6 \times 3) + (6 \times 7) = 6 \times (3 + 7) = 6 \times 10$.

CCSS: 3.MD.7.c

Practice

Directions: For questions 1 through 3, use the distributive property and the rectangle area models to help you complete the number sentences.

1.

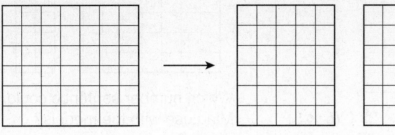

$6 \times 7 = 6 \times ($ _____ $+$ _____ $)$

$= (6 \times$ _____ $) + (6 \times$ _____ $)$

2.

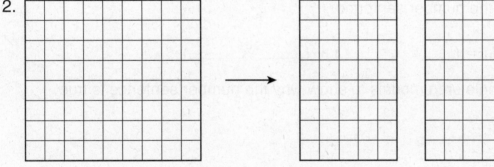

_____ $\times$ _____ $=$ _____ $\times ($ _____ $+$ _____ $)$

$= ($ _____ $\times$ _____ $) + ($ _____ $\times$ _____ $)$

3.

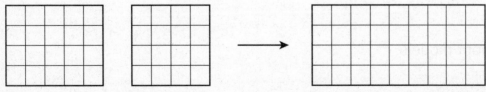

$($ _____ $\times$ _____ $) + ($ _____ $\times$ _____ $) =$ _____ $\times ($ _____ $+$ _____ $)$

$=$ _____ $\times$ _____

171

4. Which number sentence could be used with the model to show breaking the model apart?

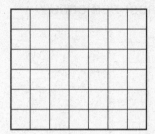

A. $6 \times (2 + 5) = (6 \times 2) + (6 \times 5)$

B. $6 \times 2 \times 5 = (6 \times 2) \times (6 \times 5)$

C. $6 + 2 \times 5 = (6 + 2) \times (6 + 5)$

D. $6 + 2 \times 5 = (6 + 2) + (6 + 5)$

5. Mina drew these two rectangle area models when explaining the distributive property to Adam.

Which number sentence could Mina use with the models?

A. $(5 \times 7) + (2 \times 7) = 7 \times (5 + 2)$

B. $(5 \times 7) + (5 \times 2) = 5 \times (7 + 2)$

C. $(7 \times 5) + (7 \times 2) = 7 \times (2 + 5)$

D. $(7 \times 7) + (2 \times 2) = 9 \times (7 + 2)$

6. Complete the number sentence.

$3 \times (3 + 6) = (\underline{\hspace{1cm}} \times \underline{\hspace{1cm}}) + (\underline{\hspace{1cm}} \times \underline{\hspace{1cm}})$

Use rectangle area models to show why the number sentence is true.

Explain your models.

_____

_____

_____

CCSS: 3.MD.7.d

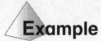

Lesson 28: Irregular Figures

You can find the area of an irregular figure by breaking it apart into rectangles or squares and finding the area of each part. The sum of the areas of the parts is the area of the irregular figure.

Example

What is the area of this figure?

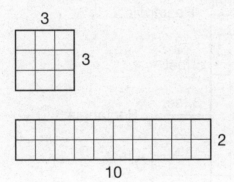

You can separate the figure into a square and a rectangle.

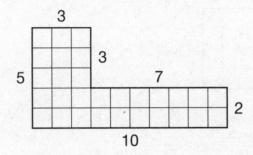

Find the area of each part.

Area of the square: $3 \times 3 = 9$
Area of the rectangle: $2 \times 10 = 20$

Find the sum of the areas.

$9 + 20 = 29$

The area of the figure is 29 square units.

⚠️ **Example**

Jason drew the following plan for a garden.

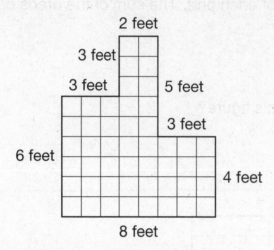

What is the area of the garden?

Separate the figure into three rectangles.

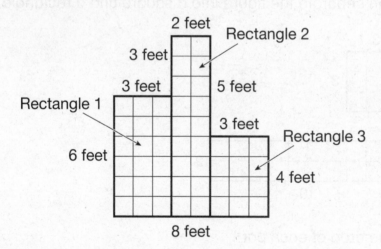

Find the area of each rectangle.

Area of Rectangle 1: $6 \times 3 = 18$
Area of Rectangle 2: $9 \times 2 = 18$
Area of Rectangle 3: $4 \times 3 = 12$

Find the sum of the areas of the three rectangles.
$18 + 18 + 12 = 48$

The area of the garden is 48 square feet.

 TIP: There is often more than one way to break apart an irregular figure.

CCSS: 3.MD.7.d

Practice

Directions: For questions 1 through 3, find the area of each irregular figure.

1.

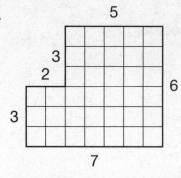

The area is _____ square units.

2.

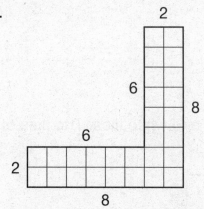

The area is _____ square units.

3.

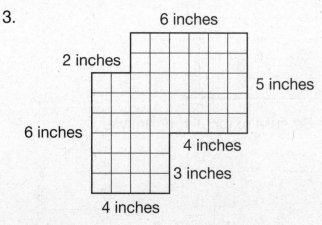

The area is _____ square inches.

4. The figure shows Mr. Hom's backyard.

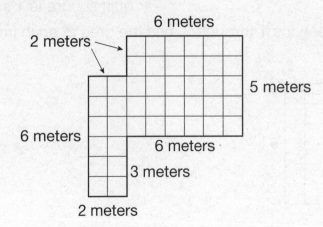

What is the area of the backyard? Be sure to label your answer.

_____

5. Christopher is making a mosaic using gray tiles and white tiles. The tiles are 1-inch squares.

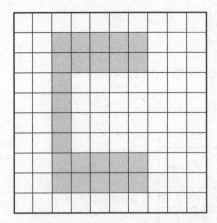

What is the area of the gray tiles? Be sure to label your answer.

_____

CCSS: 3.MD.7.d

6. What is the area of the figure shown below?

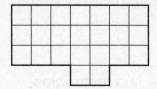

   A. 21 square units

   B. 23 square units

   C. 27 square units

   D. 28 square units

7. Mrs. Ellis is sewing a quilt. Each quilt square is 1 square foot.

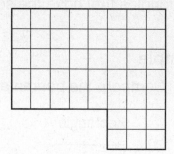

What is the area of the quilt?

   A. 21 square feet

   B. 25 square feet

   C. 46 square feet

   D. 56 square feet

8. Arabella drew this plan for a tiled patio.

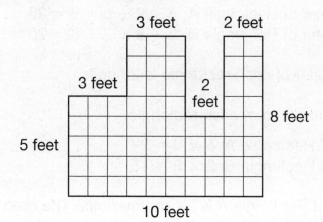

What is the area of the patio? Be sure to label your answer.

_____

Explain how you found your answer.

_____

_____

_____

CCSS: 3.MD.8

Lesson 29: Relating Perimeter and Area

Two rectangles that have the same perimeter can have different areas.

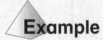

Example

Compare the perimeters and areas of these two rectangles.

Rectangle A

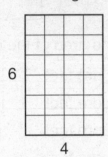

6

4

Rectangle B

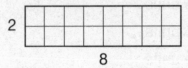

2

8

First, find the perimeter of each rectangle.

Perimeter of Rectangle A: 6 + 4 + 6 + 4 = 20
Perimeter of Rectangle B: 2 + 8 + 2 + 8 = 20

The perimeter of each rectangle is 20 units.

Next, find the area of each rectangle.

Area of Rectangle A: $6 \times 4 = 24$
Area of Rectangle B: $2 \times 8 = 16$

The area of Rectangle A is 24 square units. The area of Rectangle B is 16 square units.

The two rectangles have the same perimeter, but their areas are different.

CCSS: 3.MD.8

Two rectangles that have the same area can have different perimeters.

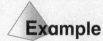 **Example**

Jenna and Darrel each turned in a design for a play area. Compare the perimeters and the areas of these two designs.

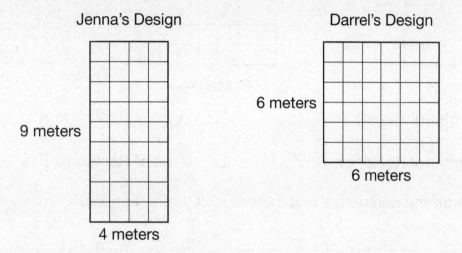

Jenna's Design

Darrel's Design

9 meters

4 meters

6 meters

6 meters

First, find the perimeter of each design.

Perimeter of Jenna's Design: 9 + 4 + 9 + 4 = 26
Perimeter of Darrel's Design: 6 + 6 + 6 + 6 = 24

The perimeter of Jenna's design is 26 meters. The perimeter of Darrel's design is 24 meters.

Next, find the area of each design.

Area of Jenna's Design: 9 × 4 = 36
Area of Darrel's Design: 6 × 6 = 36

The area of each design is 36 square meters.

The two designs have different perimeters, but the areas are the same.

Practice

Directions: For questions 1 and 2, find the perimeter and the area of each rectangle. Then compare.

1.

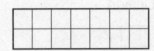

Rectangle A   Rectangle B

Perimeter of Rectangle A: _____   Area of Rectangle A: _____

Perimeter of Rectangle B: _____   Area of Rectangle B: _____

Compare the perimeters and the areas of the rectangles.

_____

_____

2.

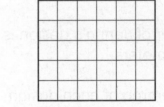

Rectangle A   Rectangle B

Perimeter of Rectangle A: _____   Area of Rectangle A: _____

Perimeter of Rectangle B: _____   Area of Rectangle B: _____

Compare the perimeters and the areas of the rectangles.

_____

_____

CCSS: 3.MD.8

3. Which statement about the perimeters is true?

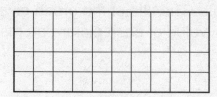 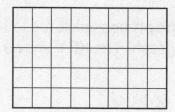

Rectangle A   Rectangle B

A. The perimeter of Rectangle A is 24 units.

B. The perimeter of Rectangle B is 28 units.

C. The perimeters of Rectangles A and B are the same.

D. The perimeter of Rectangle A is greater than the perimeter of Rectangle B.

Directions: For questions 4 and 5, use the information and diagrams below.

Carl and Tatiana each made a tile tray using 1-inch square tiles.

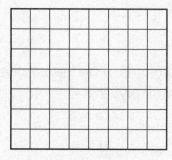 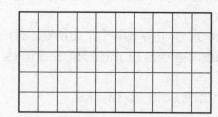

Carl's tray   Tatiana's tray

4. Which tray has the greater area? _____

Explain how you found your answer.

_____

_____

5. Which tray has the greater perimeter? _____

Explain how you found your answer.

_____

_____

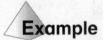

Lesson 30: Picture Graphs

A **picture graph** uses pictures to show data. The title of the graph gives the main idea of the graph. The KEY shows you the value of each picture on the picture graph.

Example

Harry kept track of how many of each menu item he sold at his restaurant one day. His results are shown in the following picture graph.

Items Sold at Harry's Hot Dogs

hot dogs	🥤🥤🥤🥤🥤🥤🥤
corn dogs	🥤🥤🥤🥤
drinks	🥤🥤🥤🥤🥤🥤🥤🥤🥤

KEY

🥤 = 5 sold

How many drinks did Harry sell?

Look at the row for drinks. There are 9 glasses. The key shows that each glass stands for 5 drinks sold, so multiply: 9 × 5 = 45.

Harry sold 45 drinks.

You can use information from a picture graph to solve problems.

How many more hot dogs were sold than corn dogs at Harry's Hot Dogs?

Look at the row for hot dogs. There are 7 glasses. Each glass stands for 5 hot dogs sold, so multiply: 7 × 5 = 35.
35 hot dogs were sold.

Look at the row for corn dogs. There are 4 glasses. Each glass stands for 5 corn dogs sold, so multiply: 4 × 5 = 20.
20 corn dogs were sold.

Subtract to find the difference: 35 − 20 = 15.

Harry sold 15 more hot dogs than corn dogs.

CCSS: 3.MD.3

One way to collect and organize data is in a frequency table. Then, you can use the table to make a graph to display the data.

 **Example**

Robert recorded the color of his classmates' eyes in a frequency table.

Students' Eye Color

Color	Number of Students	
Brown	꙰꙰꙰꙰꙰꙰꙰꙰꙰꙰꙰꙰	12
Blue	꙰꙰꙰꙰꙰꙰꙰꙰	8
Green	‖‖‖‖	4
Other	꙰꙰꙰꙰꙰꙰	6

Make a picture graph of the data.

Decide on the key for the picture graph, and choose a symbol.
The information is about the number of students. The symbol can be a picture of a student. Since all the numbers are even, let each symbol stand for 2 students.

Find the number of symbols to use for each eye color. For each eye color, divide the number of students by 2.

Fill in the picture graph. Include a title and the KEY.

Students' Eye Color

Brown	👤 👤 👤 👤 👤 👤
Blue	👤 👤 👤 👤
Green	👤 👤
Other	👤 👤 👤

KEY
👤 = 2 students

Practice

Directions: Use the following information to answer questions 1 through 6.

Three students were in charge of sending out invitations for a class party. The following picture graph shows the number of invitations that Ty, Ann, and Zoe sent to their classmates.

**Invitations Sent to
Classmates**

Ty	✉ ✉ ✉
Ann	✉ ✉
Zoe	✉ ✉ ✉ ✉

KEY
✉ = 2 invitations sent

1. How many invitations did Ann send? _____

2. Who sent the most invitations? _____

3. How many invitations did Ty send? _____

4. How many invitations were sent in all? _____

5. How many more invitations did Zoe send than Ty? _____

6. How many invitations did Zoe and Ann send combined? _____

   Explain how you found your answer.

   _____

   _____

   _____

CCSS: 3.MD.3

Directions: Use the following information to answer questions 7 through 10.

This frequency table shows how many milk shakes, fruit smoothies, turkey wraps, and hamburgers were sold during one month at Hamburger Bob's.

Items Sold at Hamburger Bob's During One Month

Item	Number Sold	
Milk Shakes	⟋⟋⟋⟋⟋ ⟋⟋⟋⟋⟋ ⟋⟋⟋⟋⟋	15
Fruit Smoothies	⟋⟋⟋⟋⟋ ⟋⟋⟋⟋⟋ ⟋⟋⟋⟋⟋ ⟋⟋⟋⟋⟋	20
Turkey Wraps	⟋⟋⟋⟋⟋ ⟋⟋⟋⟋⟋ ⟋⟋⟋⟋⟋ ⟋⟋⟋⟋⟋ ⟋⟋⟋⟋⟋ ⟋⟋⟋⟋⟋	30
Hamburgers	⟋⟋⟋⟋⟋ ⟋⟋⟋⟋⟋ ⟋⟋⟋⟋⟋ ⟋⟋⟋⟋⟋ ⟋⟋⟋⟋⟋ ⟋⟋⟋⟋⟋ ⟋⟋⟋⟋⟋ ⟋⟋⟋⟋⟋	40

KEY
| = 1
⟋⟋⟋⟋⟋ = 5

7. Make a picture graph of the data. Use the symbol found in the KEY.

Items Sold at Hamburger Bob's During One Month

Milk Shakes	
Fruit Smoothies	
Turkey Wraps	
Hamburgers	

KEY

△ = 5 sold

8. How many fruit smoothies were sold? _____

9. How many more fruit smoothies were sold than milk shakes? _____

10. How many turkey wraps and hamburgers were sold in all? _____

Directions: Use the picture graph to answer questions 11 through 14.

The picture graph below shows the number of laps members of the swim club swam at their last practice.

Swim Club

Student	Number of Laps
Lisa	👓 👓
Tia	👓 👓 👓 👓 👓
Lex	👓 👓 👓 👓
Rick	👓 👓 👓

KEY
👓 = 5 laps

11. Who swam the greatest number of laps?

    A. Lisa

    B. Tia

    C. Lex

    D. Rick

12. How many more laps did Lex swim than Lisa?

    A. 2

    B. 4

    C. 5

    D. 10

13. Tracie joined the swim team. She was able to swim 5 laps. You want to add Tracie's data to the Swim Club picture graph. How many symbols should you use?

    _____

14. Eric looked at the Swim Club picture graph. He said that the graph shows that Rick swam 1 more lap than Lisa. Do you agree?

    _____

    Explain your answer.

    _____

    _____

    _____

CCSS: 3.MD.3

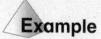

Lesson 31: Bar Graphs

A **bar graph** uses bars to compare information. The bars can go up and down or left and right. Be sure to check the title and labels on the graph.

Example

The third graders at Shady Lane Elementary School hung bird feeders outside at their school. They kept track of how many blackbirds came each month of the school year.

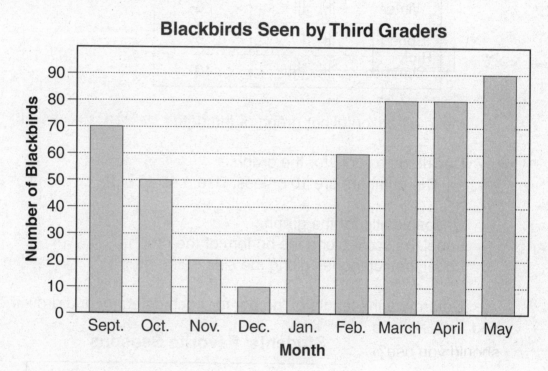

How many blackbirds came to the feeders in March?

Find the bar for March. Follow the bar to the top. Look across to the left to find the number that meets the top of the bar. The number 80 lines up with the top of the bar for March.

In March, 80 blackbirds came to the bird feeders.

During which month did the fewest number of blackbirds come to the bird feeders?

Find the shortest bar, and read the name of the month. December is the month with the shortest bar.

December had the fewest number of blackbirds at the bird feeders.

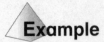

Example

The frequency table shows the favorite seasons of students in Ms. Wilson's third grade class.

Students' Favorite Seasons

Season		Number of Students
Fall	||	2
Winter	ЖЖ |||	8
Spring	ЖЖ |	6
Summer	ЖЖ ЖЖ	10

Make a horizontal bar graph of the data.

Decide on a scale for the graph.
    The numbers are 10 or less. Use a scale of 2.

Choose a title for the graph.
Label the scale along the bottom of the graph.
Label the categories along the side of the graph.

Determine the length of the bar for each category and draw the graph.

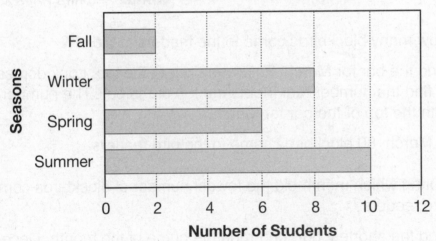

CCSS: 3.MD.3

Practice

Directions: Use the following information to answer questions 1 through 4.

Marcus and his dad went fishing on Falls Lake last week. The bar graph shows how many white bass they caught each day.

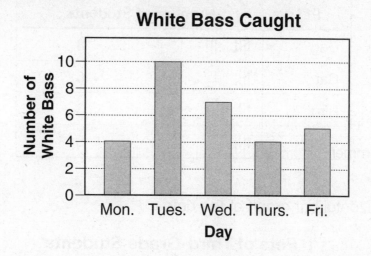

White Bass Caught

1. How many white bass did they catch on Wednesday? _____

2. How many more white bass did they catch on Friday than on Thursday?

   _____

3. On which two days did they catch the least number of white bass?

   _____

4. How many more bass did they catch on Tuesday than on Thursday and Friday combined?

   _____

   Explain how you found your answer.

   _____

   _____

   _____

Directions: Use the following information to answer questions 5 through 8.

This frequency table shows the pets of students in the third grade at Hudson Elementary School.

Pets of Third-Grade Students

Pet	Number of Students	
Dog	ⅢⅠ ‖‖	8
Cat	ⅢⅠ Ⅰ	6
Bird	‖‖	3
Other	ⅢⅠ	5

5. Make a horizontal bar graph of the data.

Pets of Third-Grade Students

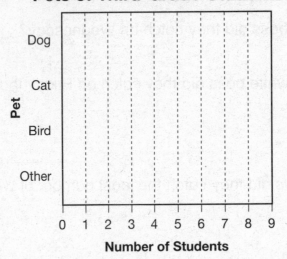

6. Which pet do the greatest number of students have? _____

7. How many more students have cats than have birds? _____

8. How many pets do the students in the third grade have in all?

_____

CCSS: 3.MD.3

Directions: Use the bar graph to answer questions 9 through 13.

Lucy asked each of her classmates what his or her favorite snack is. Her results are shown in this bar graph.

9. How many more classmates chose cheese than popcorn?

    A. 1

    B. 2

    C. 4

    D. 10

10. Which snack did the least number of students choose?

    A. cheese

    B. nuts

    C. popcorn

    D. fruit

11. How many classmates chose fruit or nuts? _____

12. Which two snacks were chosen by the same number of classmates?

    _____

13. Lucy forgot to include the data for classmates that chose veggies and dip. There were 12 classmates that chose veggies and dip. How would you show that information on the bar graph?

    _____

    _____

    _____

Lesson 32: Line Plots

A **line plot** uses marks to show the number of times that each value or result occurs.

Example

In their last ten football seasons, Jeb's team won 7, 9, 7, 6, 8, 13, 7, 10, 9, and 4 games. The following line plot shows the data.

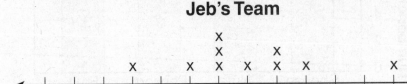

Jeb's Team

Number of Games Won in a Season

During how many seasons did Jeb's team win 9 games?

Each X on the line plot shows a season in which Jeb's team won that number of games. Look above the 9. There are 2 Xs.

Jeb's team won 9 games during 2 seasons.

During how many seasons did Jeb's team win 8 or more games?

Find the number of Xs above 8 and above each number greater than 8. Notice that there are no Xs above 11, 12, or 14. That means that during no season did Jeb's team win 11, 12, or 14 games.

8 games → 1 X
9 games → 2 Xs
10 games → 1 X
13 games → 1 X

Find the sum of the number of Xs above 8 and above each number greater than 8.

1 + 2 + 1 + 1 = 5 Xs

Jeb's team won 8 or more games during 5 seasons.

A line plot can also show a set of measurements in fractions of an inch.

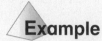 **Example**

Ricky measured the bolts that his father has in a small box. The following are the lengths of the bolts in inches.

$1\frac{2}{4}$ $2\frac{1}{4}$ $2\frac{3}{4}$ $2\frac{2}{4}$ $3\frac{1}{4}$ $2\frac{2}{4}$ $2\frac{1}{4}$ $2\frac{1}{4}$ $2\frac{3}{4}$ 2 $2\frac{1}{4}$ $2\frac{1}{4}$

Make a line plot to shows the data.

First, find the least number and the greatest number you need to show.

The least number is $1\frac{2}{4}$. The greatest number is $3\frac{1}{4}$.

Draw a number line that includes those numbers.

Make the number line from 1 to $3\frac{1}{4}$ and make a tick mark at every quarter inch.

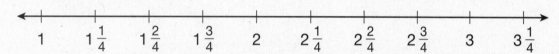

Place an X above the number that stands for the length of each bolt.

Lengths of Bolts Measured by Ricky

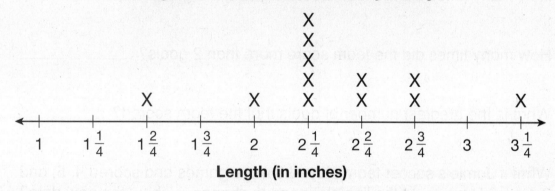

Length (in inches)

How many bolts are greater than 2 inches in length?

To find the number of bolts that are greater than 2 inches, count the Xs that are at $2\frac{1}{4}$ or greater.

$5 + 2 + 2 + 1 = 10$ Xs

Ten bolts are greater than 2 inches in length.

Practice

Directions: Use the line plot to answer questions 1 through 6.

In the last ten games, Jamie's soccer team scored 1, 1, 4, 4, 1, 2, 0, 3, 1, and 1 goals. The following line plot shows the data.

Jamie's Soccer Team

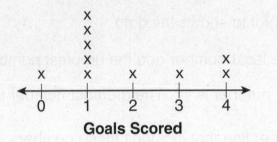

Goals Scored

1. What is the least number of goals that Jamie's soccer team scored in a game?

    _____

2. How many times did the team score 4 goals in a game? _____

3. How many times did the team score 3 goals in a game? _____

4. How many times did the team score more than 2 goals? _____

5. What is the greatest number of goals that the team scored? _____

6. What if Jamie's soccer team played 3 more games and scored 4, 5, and 2 goals? How would the line plot need to change to show the new data?

    _____

    _____

    _____

Directions: For questions 7 through 11, use the following bars.

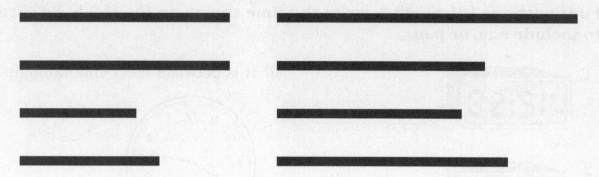

7. What is the measure of each bar to the nearest $\frac{1}{4}$ inch?

_____

8. Complete the line plot to show the data on bar lengths.

Lengths of Bars

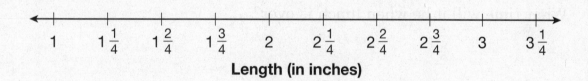

Length (in inches)

9. How many bars measure $2\frac{2}{4}$ inches? _____

10. How many bars measure less than 2 inches? _____

11. Explain how you created the line plot for Lengths of Bars.

_____

_____

_____

Unit 4 Practice Test

For questions 1 through 3, write the time shown on the clock. Be sure to include a.m. or p.m.

1.

_____

3. It is between noon and midnight.

_____

2.

_____

4. It is 12:10 p.m. Lunch will be over in 15 minutes.

What time will it be when lunch is over? _____

5. Mr. Evans had the oven on for a total of 50 minutes. The oven was preheating for 7 minutes. A pie was baking in the oven during the rest of the time. How long was the pie baking? _____

6. Milk is delivered to a store in crates. Each crate holds 6 bottles of milk. Each bottle contains 2 liters. How many liters of milk are in each crate? _____

7. For a snack, Juan has two crackers which are a total of 28 grams and two pieces of cheese which are a total of 56 grams. How many grams is Juan's snack? _____

8. Find the perimeter of the figure.

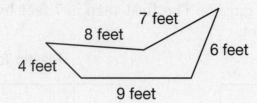

The perimeter is _____.

9. Find the area of the figure in square units.

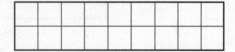

The area is _____ square units.

10. Find the area of the figure.

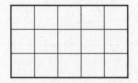 = 1 square centimeter

The area is _____ square _____.

11. Alana's patio has a tile floor. Each tile is 1 square foot.

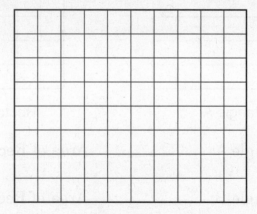

Which multiplication fact could you use to find the area of the patio?

_____

12. Bobby needs 56 square feet of plastic tarp. There are two rectangular pieces of tarp in the garage. The first tarp is 7 feet by 8 feet. The second tarp is 6 feet by 9 feet.

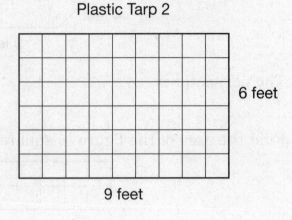

Plastic Tarp 1   Plastic Tarp 2

7 feet   6 feet

8 feet   9 feet

Which tarp should Bobby use? _____

Explain how you found your answer.

_____

_____

_____

13. Find the perimeter and the area of each rectangle. Then compare.

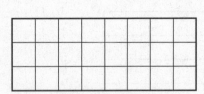   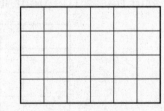

Rectangle A   Rectangle B

Perimeter of Rectangle A: _____   Area of Rectangle A: _____

Perimeter of Rectangle B: _____   Area of Rectangle B: _____

Compare the perimeters and the areas of the rectangles.

_____

_____

14. Which is the best estimate for the capacity of a sauce pan?

A. 1 liter

B. 10 liters

C. 100 liters

D. 1000 liters

15. Which is the best estimate for the mass of a large puppy?

A. 2 kilograms

B. 10 kilograms

C. 26 kilograms

D. 50 kilograms

Use the bar graph to answer questions 16 and 17.

Casey was training for a race. The bar graph below shows the number of miles he ran each week during a four-week period.

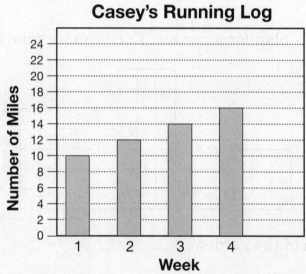

16. How many more miles did Casey run in week 4 than in week 1?

A. 3

B. 6

C. 10

D. 16

17. What is the total number of miles that Casey ran during the four-week period?

A. 16

B. 40

C. 52

D. 56

18. Which number sentence could be used with the model to show the distributive property?

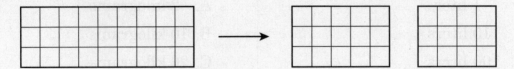

   A. $3 \times (5 \times 4) = (3 \times 5) \times (3 \times 4)$

   B. $3 + (5 + 4) = (3 + 5) + (3 + 4)$

   C. $3 + (5 \times 4) = (3 + 5) \times (3 + 4)$

   D. $3 \times (5 + 4) = (3 \times 5) + (3 \times 4)$

19. Frankie drew this plan for a garden. Each square represents 1 square foot.

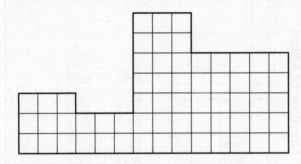

What is the area of the garden? _____

Show your work.

Explain how you found your answer.

_____

_____

_____

20. How are the units used to describe area different from the units used to describe length?

_____

_____

21. The frequency table shows the tourist attractions that students in Ms. Carter's third grade class would most like to visit.

Complete the frequency table.

Tourist Attractions

Tourist Attraction	Number of Students	
Niagara Falls	ⵌ ⵌ II	
The Grand Canyon	ⵌ I	
Statue of Liberty	ⵌ III	
The White House	IIII	

Make a picture graph of the data. Use the symbol found in the KEY in your picture graph.

Tourist Attractions

Niagara Falls	
The Grand Canyon	
Statue of Liberty	
The White House	

KEY
🯅 = 2 students

22. Alicia has a box of used pencils.

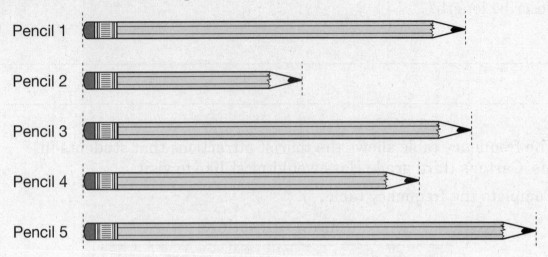

Part A
Measure each pencil to the nearest $\frac{1}{2}$ inch.

Pencil 1: _____   Pencil 4: _____

Pencil 2: _____   Pencil 5: _____

Pencil 3: _____

Part B
Complete the line plot to show the data on pencil lengths.

Lengths of Pencils

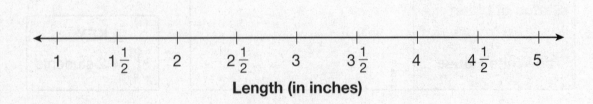

Length (in inches)

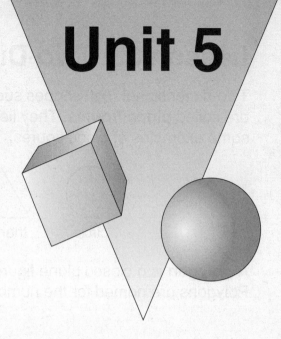

Unit 5

Geometry

There are many types of polygons in the objects around you. Many walls, doors, windows, books, and sheets of paper are shaped like rectangles or squares. Many of the figures you see daily have four sides, but there are also figures with more sides. For example, a stop sign has 8 straight sides. Engineers and architects pay careful attention to shapes. Different shapes have different properties. These properties are important in building bridges and skyscrapers, in making airplanes fly, and in making engines run efficiently.

In this unit, you will describe and classify two-dimensional shapes. You will also break shapes into parts with equal areas.

In This Unit

Two-Dimensional Shapes

Quadrilaterals

Breaking Shapes into Equal Areas

Lesson 33: Two-Dimensional Shapes

Two-dimensional (flat) shapes such as circles, triangles, squares, and rectangles are called **plane figures**. They lie on a flat surface, or plane. The following are some examples of plane figures.

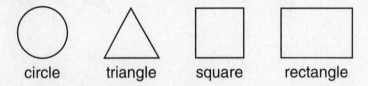

circle   triangle   square   rectangle

A **polygon** is a closed plane figure made up of three or more straight sides. Polygons are named for the number of sides and vertices they have.

side → vertex

The following chart shows the names of some kinds of polygons.

Polygon	Name	Number of Sides and Vertices
	triangle	3
	quadrilateral	4
	pentagon	5
	hexagon	6
	octagon	8

 TIP: Squares and **rectangles** are quadrilaterals.

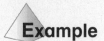**Example**

Jason drew this figure.

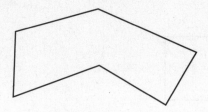

Is it a hexagon?

A hexagon has 6 straight sides. Count the sides.
The figure has 6 sides.

Yes, the figure is a hexagon.

A **regular polygon** has all sides the same length. An irregular polygon has sides of different lengths.

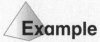**Example**

Which figure is a regular pentagon?

Figure 1

Figure 2

Both figures have 5 sides, so both figures are pentagons.
The first figure has all sides that are the same length. The second figure has sides of different lengths.

Figure 1 is a regular pentagon.

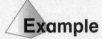

Example

Compare the two figures. How are the figures alike? How are they different?

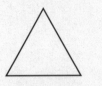

Figure 1

Figure 2

First, look at the figures.
They are both polygons. Figure 1 has 3 equal sides. Figure 2 has 4 equal sides.

The figures are alike in that they both are polygons and have sides that are equal in length.
The figures are different in that they have a different number of sides.

Practice

1. Shade the two-dimensional shapes that are polygons.

2. Shade the two-dimensional shape that is **not** a polygon.

What is the name of the two-dimensional shape that you shaded?

_____

Why is this two-dimensional shape **not** a polygon?

_____

_____

CCSS: 3.G.1

3. Draw a line from each real-life object to the name of the polygon with the same shape.

triangle   pentagon   hexagon   quadrilateral   octagon

Directions: For questions 4 through 6, write the name of the polygon with the given number of sides and vertices.

4. 3 sides and 3 vertices _____

5. 8 sides and 8 vertices _____

6. 6 sides and 6 vertices _____

7. How many sides and vertices does a rectangle have? _____

   What is another word that can be used to describe a rectangle?

   _____

Directions: For questions 8 and 9, draw an example of the given polygon in the space provided.

8. pentagon

9. quadrilateral

10. Which polygon is **not** a triangle?

A.

B.

C.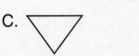

D.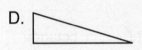

11. Which polygon is a quadrilateral?

A.

B.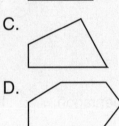

C.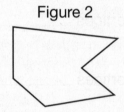

D.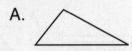

12. Write the name of each figure.

Figure 1

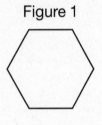

Figure 2

_____   _____

How are the figures alike? How are they different?

_____

_____

_____

CCSS: 3.G.1

Lesson 34: Quadrilaterals

Quadrilaterals with sides that are all the same length have special names. Other quadrilaterals with sides that meet to form square corners also have special names.

Remember, a quadrilateral is a polygon with 4 sides and 4 vertices. When the sides of a polygon meet to form square corners, this can be shown using the square corner symbol.

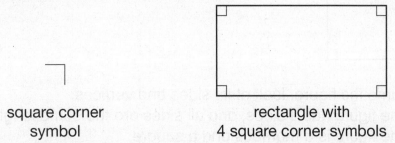

square corner
symbol

rectangle with
4 square corner symbols

Rectangles

A **rectangle** has four square corners formed by its sides. All the sides do **not** have to be the same length.

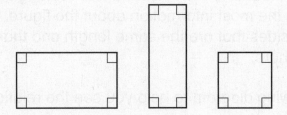

Rhombuses

A **rhombus** has four sides with the same length.

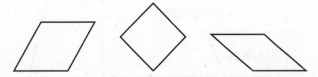

Squares

A **square** has four square corners and four sides with the same length.

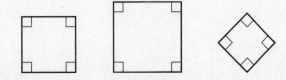

209

You usually name a figure by the group that gives the most information about its shape. However, a figure can be called by the name of any group it belongs to.

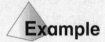

Example

Name the figure. What are all the quadrilateral groups to which the figure belongs?

To name the figure, look at its sides and vertices.
The figure has 4 sides, and all sides are the same length.
The figure is a rhombus and a square.

The figure has 4 square corners.
The figure is a rectangle and a square.

The figure is a square, a rhombus, and a rectangle. Naming it as a square gives the most information about the figure. It tells us that the figure has 4 sides that are the same length and that the figure has 4 square corners.

You can use the following diagram to help you see the relationships among the quadrilaterals.

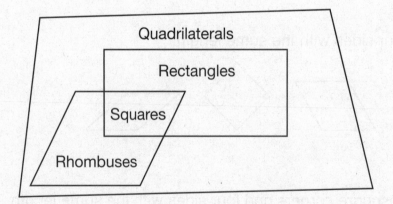

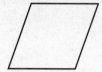

Practice

Directions: For questions 1 through 5, write the name or names (rhombus, rectangle, square, quadrilateral) that describe each figure. Some of the figures can be described by more than one name.

1.

_____

2.

_____

3.

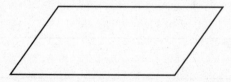

_____

4.

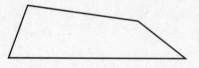

_____

5.

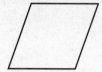

_____

6. Which of the following
   quadrilaterals is **not** a rectangle?

A.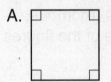

B.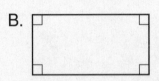

C.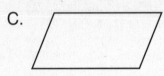

D.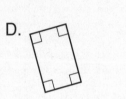

7. Which of the following
   quadrilaterals is a rhombus?

A.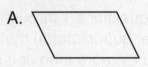

B.

C.

D.

8. Isaac says that all rectangles are squares. Flor says that all squares are rectangles.

Who is correct?

_____

Explain your answer.

_____

_____

_____

_____

CCSS: 3.G.2

Lesson 35: Breaking Shapes into Equal Areas

You can break a shape into parts with equal areas. The area of each part is a fraction of the area of the whole shape. The number of equal parts is the denominator of the fraction.

$$\text{Area of one part of a figure} = \frac{1}{\text{number of parts}} \text{ of the whole figure}$$

So, if a figure is broken into 3 equal parts, then the area of each part is $\frac{1}{3}$ of the area of the whole figure.

Example

A triangle is broken into 2 equal parts. Name the area of each part as a fraction of the area of the whole triangle.

First, find the denominator of the fraction.
The triangle is in 2 parts. The denominator of the fraction is 2.

We want to name the area of one part, so the numerator is 1.
The fraction is $\frac{1}{2}$.

The area of each part of the triangle is $\frac{1}{2}$ the area of the whole triangle.

Example

Emilio has a paper hexagon. He cuts the hexagon into 6 equal parts. Describe the area of each part as a fraction of the area of the hexagon.

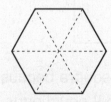

The area of each part of the hexagon is $\frac{1}{6}$ the area of the whole hexagon.

Example

Find two ways to divide a square into four equal parts.

One way to separate the square into 4 equal parts is to fold it in half lengthwise and then in half again. When the figure is opened out, there are 4 equal parts.

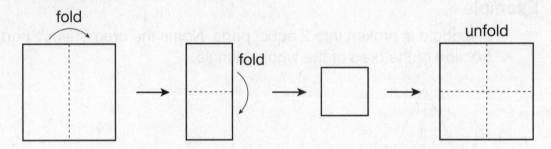

Another way is to fold the square along its diagonal and then fold it in half again. When the figure is opened out, there are 4 equal parts.

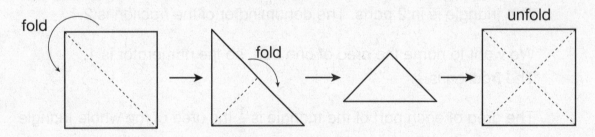

What fraction of the large square is the area of each part?

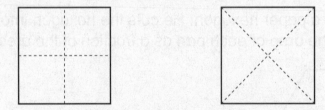

Each part represents $\frac{1}{4}$ of the area of the large square because each part is formed by breaking the large square into 4 equal parts.

Notice that the square is divided differently and the parts have a different shape, but the area of each part is the same: $\frac{1}{4}$.

CCSS: 3.G.2

Practice

Directions: For questions 1 through 6, write a fraction that describes the area of one part of each figure.

1.

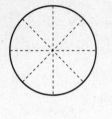

_____

2.

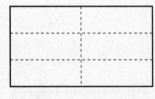

_____

3.

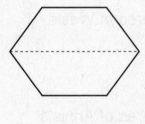

_____

4.

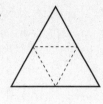

_____

5.

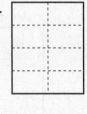

_____

6.

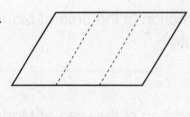

_____

7. Which figure is **not** divided into 6 equal areas?

A.

B.

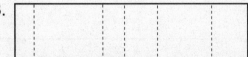

C.

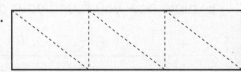

D.

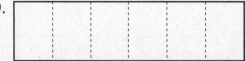

8. Which figure is divided into 4 equal parts?

A.

C.

B.

D.

9. Marlene drew a large rectangle. Wesley and Annie each drew smaller shaded rectangles.

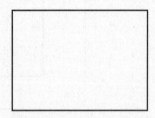

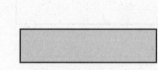

Wesley's rectangle   Marlene's rectangle   Annie's rectangle

What fraction of the area of Marlene's rectangle is the area of Wesley's rectangle?

_____

What fraction of the area of Marlene's rectangle is the area of Annie's rectangle?

_____

Explain how you found your answers.

_____

_____

_____

Unit 5 Practice Test

For questions 1 and 2, write the name of the polygon with the given number of sides and vertices.

1. 3 sides and 3 vertices _____

2. 5 sides and 5 vertices _____

3. Write the name or names (rhombus, rectangle, square, quadrilateral) that describe the figure.

_____

4. Draw an example of a hexagon in the space provided.

5. Write the name of each figure.

Figure 1

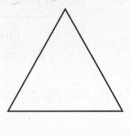

_____

Figure 2

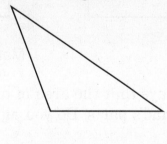

_____

How are the figures alike? How are they different?

_____

_____

_____

6. Which figure is **not** divided into 8 equal areas?

A.

B.

C.

D.

7. Which figure has the same number of sides as a square?

A. hexagon

B. pentagon

C. rectangle

D. triangle

8. Which statement is true?

A. All quadrilaterals have 4 square corners.

B. All rectangles have 4 equal sides.

C. All rhombuses have 4 square corners.

D. All squares have 4 square corners.

9. Matthew and Ahmad each separated a rectangle into 2 equal parts.

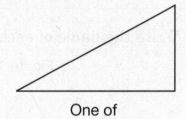

One of
Matthew's parts

One of
Ahmad's parts

Matthew says that the area of one of his parts is the same as the area of one of Ahmad's parts. Do you agree?

_____

Explain your answer.

10. Which of these polygons is a pentagon?

A.

B.

C.

D.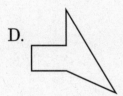

11. Which of the following quadrilaterals is not a rhombus?

A.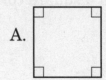

B.

C.

D.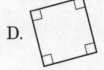

12. Lester says that all rectangles are quadrilaterals and all quadrilaterals are rectangles. Do you agree with Lester?

_____

Explain your answer.

_____

_____

_____

Draw a picture to support your answer.

13. Use what you know about plane figures and area to answer the questions.

Part A
Draw a figure with 4 square corners and sides of two different lengths.

Part B
What names describe the figure that you drew?

_____

Part C
Draw lines to break apart your figure into 6 equal parts. What fraction of the area of the whole figure is the area of each part?

_____

Part D
Draw another figure like the one you drew in Part A. Draw lines to break apart your figure into 6 equal parts. Divide the figure in a different way from the way you did above.

What fraction of the area of the whole figure is the area of each part?

_____

Math Tool: Number Lines

Math Tool: Fraction Circles

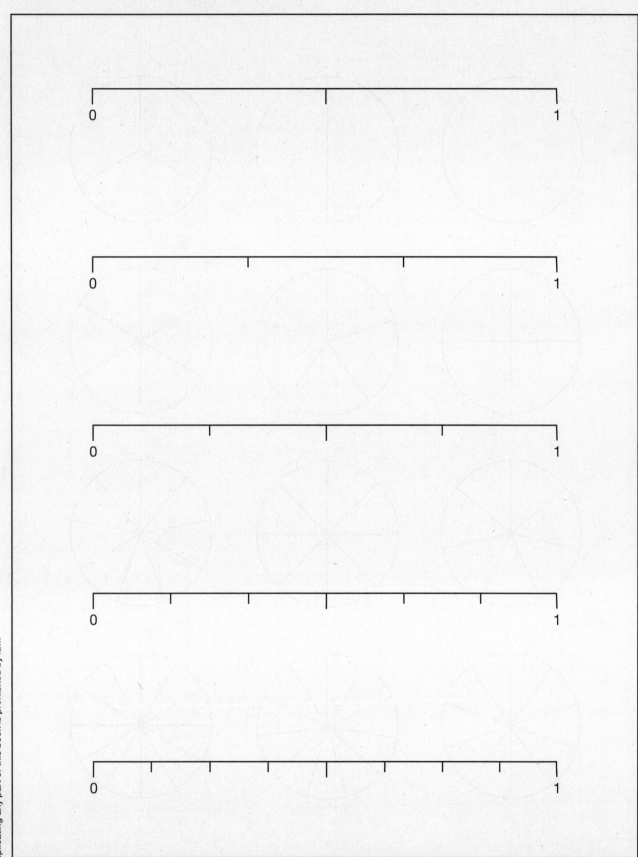

Math Tool: Fraction Circles

Math Tool: Fraction Strips

Math Tool: Grid Paper

Math Tool: Fraction Strips